This book is dedicated to
the Clayoquot,
the Walbran,
the Carmanah,
and all the rainforests of this planet.
Their defenders are
the authors
of this book.

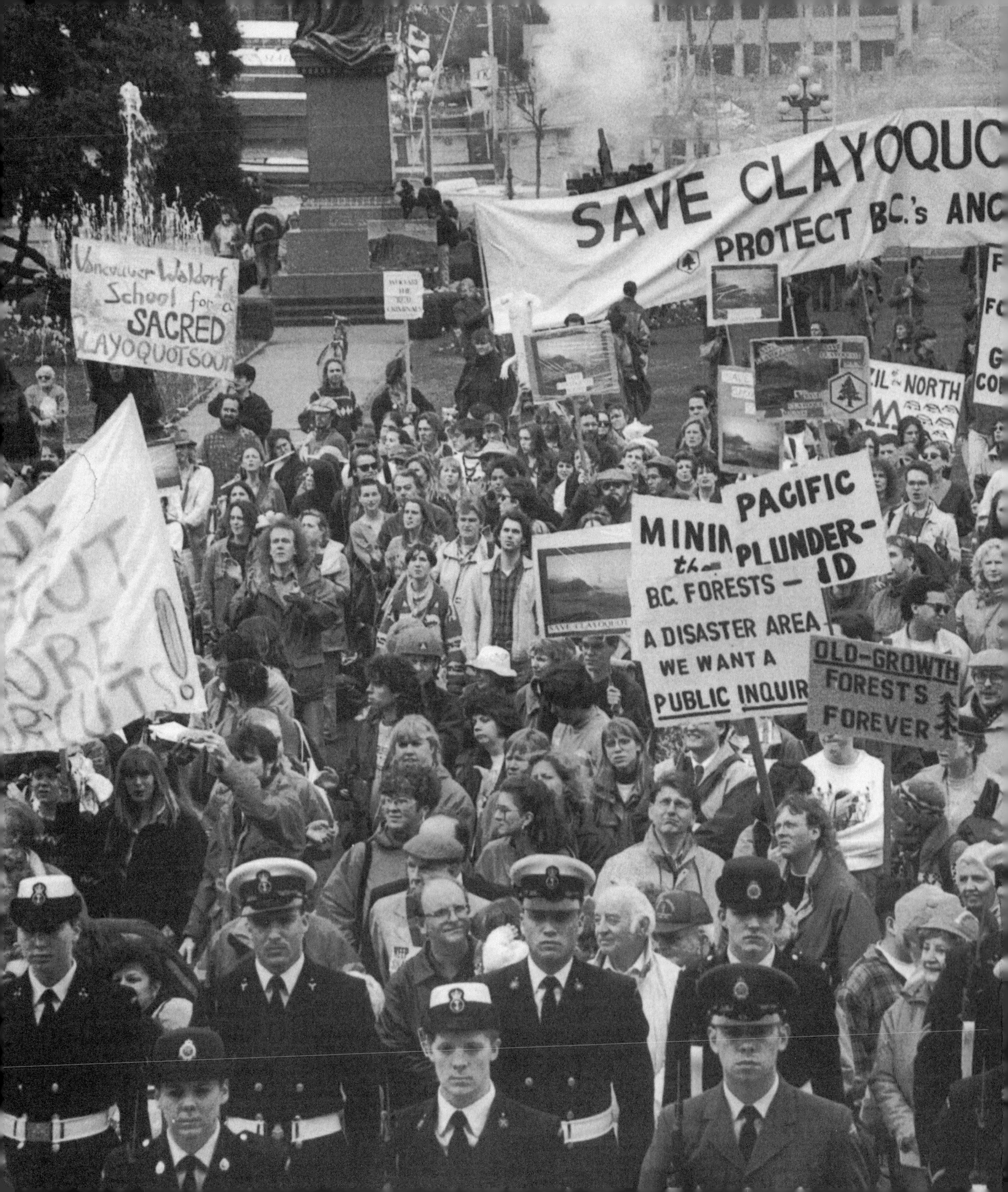
SAVE CLAYOQUO
PROTECT B.C.'s ANC
Vancouver Waldorf School for a SACRED CLAYOQUOT SOUN
PACIFIC PLUNDER-ID
MINI
B.C. FORESTS — A DISASTER AREA
WE WANT A PUBLIC INQUIR
OLD-GROWTH FORESTS FOREVER
ZIL of the NORTH

CLAYOQUOT MASS TRIALS

DEFENDING • THE • RAINFOREST

Edited by Ron MacIsaac
and Anne Champagne

Portraits of Clayoquot Protectors
by Andy Sinats

NEW SOCIETY PUBLISHERS
Philadelphia, PA   Gabriola Island, B.C.

Canadian Cataloguing in Publication Data
Main entry under title:
Clayoquot mass trials

Includes bibliographical references.
ISBN 1-55092-252-1 (bound). -- ISBN 1-55092-253-X (pbk.)
1. Clayoquot Peace Camp, Tofino, B.C., 1993. 2. Protest movements -- British Columbia. 3. Trials (Contempt of legislative bodies) -- British Columbia. 4. Clayoquot Sound Region (B.C.) -- Environmental conditions. 5. Environmental policy -- British Columbia. I. MacIsaac, Ron, 1925 - II. Champagne, Anne.
FC3829.9.C62C629 1994 971.1'2 C94-910883-9 F1089.V3C629 1994

Royalties from the sale of this book will be donated to the Friends of Clayoquot Sound.

Cover and book design by David Lester, Get to the Point Graphics.
Printed in Canada by Gagné Best Book Manufacturers, Quebec.

Inquiries regarding requests to reprint all or part of *Clayoquot Mass Trials* should be addressed to: New Society Publishers at the addresses below.

Canada ISBN: 1-55092-253-X (Paperback)
Canada ISBN: 1-55092-252-1 (Hardback)
USA ISBN: 0-86571-321-9 (Paperback)
USA ISBN: 0-86571-320-0 (Hardback)

To order directly from the publishers, please add $2.50 to the price of the first copy, and 75 cents for each additional copy (plus GST in Canada). Send check or money order to:
New Society Publishers,
P.O. Box 189, Gabriola Island, B.C. Canada V0R 1X0 (1-800-567-6772)
or
4527 Springfield Avenue, Philadelphia PA, U.S.A. 19143 (1-800-333-9093)

New Society Publishers is a project of the Catalyst Education Society, a nonprofit educational society in Canada, and the New Society Educational Foundation, a nonprofit, tax-exempt public foundation in the U.S. Opinions expressed in this book do not necessarily reflect positions of the Catalyst Education Society nor the New Society Educational Foundation.

TABLE OF CONTENTS

ACKNOWLEDGMENTS

This book would not have made it to the press without the efforts of a number of people who gave their time generously to the project.

Diane Burnett, Kathryn Burgess, and Barbara Lee worked on the court transcripts, and Joan Russow helped out in the initial stages of getting the manuscript into shape. Anne Champagne's thorough editorial efforts later in the process are especially appreciated.

I sincerely appreciate all of the writers and publishers who gave permission for their articles and extracts to be reprinted in this book. And special thanks are due to the photographers who freely donated their work: Adrian Dorst, Lawrence McLagan, Andy Sinats, and Chip Vinai. Photographs of over 330 of the arrested protectors were co-ordinated by Andy Sinats, and taken from his on-going project "The Big Picture" —see below. For the cartography of Clayoquot Sound, again donated freely, I thank Ian Parfitt of the Western Canada Wilderness Committee (1994).

Of the many people who contributed to the tremendous efforts surrounding the protest at Clayoquot and the trials in Victoria, Michael Williams deserves special thanks for having donated space for the Clayoquot Resource Centre in Victoria, as well as having loaned machinery, and paid for power, insurance, etc., in support of those arrested. I also extend special thanks to all those from overseas and from south of the border who have workedfor the Clayoquot, and especially to Robert Kennedy Jr. and the Natural Resources Defense Council from Washington DC who continue to bring pressure to bear on this issue.

Editorial note: unless otherwise indicated, all quotations are from the trials, and people's original language has been preserved as much as possible.

— Ron MacIsaac, Victoria, B.C., October 1994.

◆

Photographs of the arrestees are from "The Big Picture," a portrait of those arrested for civil disobedience in defense of the temperate rainforest of Clayoquot Sound. The inspiration of their protest provided the energy for this project, and it is dedicated to them. Thanks are also due to Chip Vinai, Karen Blake, Gordon Lee, Ian McAllister, Karl Sturmanis and Marlene Dayman for their help in photographing the very many individuals who comprise this portrait, to Jon Hoadley for his darkroom expertise and steadfast generosity, Gregory Hartnell of *La Rosa* for his spiritual support, and my wife, Petra, for being with me in this from first to last.

— Andy Sinats, Victoria, B.C., October 1994.

FOREWORD

Robert F. Kennedy, Jr.

Clayoquot Sound has become the flashpoint in one of the defining environmental battles of our time. In Clayoquot, the fight to save one thousand year-old cedars and hemlocks intertwines with the aboriginal peoples' struggle to control traditional lands and their economic destiny. As environmentalists and First Nations work to reconcile their respective visions, large-scale industrial logging continues to liquidate the forest for cash. Meanwhile, a provincial government charged with protecting the public interest stands paralyzed between its idealism and its own giant stake in the promise of instant profits.

The Clayoquot protectors assembled in the summer of 1993 as emissaries for the future. Sacrificing their own security, they elbowed their way to the table where the public trust was being pilfered, and demanded an accounting on behalf of coming generations. There, they linked shoulders with the First Nations whose special claims to the wealth of the land had also been postponed as the booty was divided.

First Nations and environmentalists won the first beachhead a decade ago on Meares Island by repelling MacMillan Bloedel's menacing flotilla of loggers, grapplers, and chainsaws. The ensuing legal battle produced an historic court injunction prohibiting logging on Meares until the rightful ownership of the island is established. This was the first combined victory against the logging industry's political and industrial juggernaut.

Ten years later, the Clayoquot protectors pushed the Meares Island beachhead on to the mainland of Vancouver Island. In doing so, they extended the boundaries of democracy in Canada and suffered the largest mass arrest in Canadian history, and the largest mass trial in the western world. Over nine hundred were jailed as they asserted public ownership over resources which, under Canadian law, are owned by all Canadians but, in historical practice, were treated as the personal fiefdoms of a few giant timber companies. Since neither Canadian nor provincial law gives Canadian citizens the authority to stop the trees from falling either judicially or through formal avenues of access to government decision-makers, the protectors used the only democratic tool available: civil disobedience.

In subjecting themselves to arrest, the Clayoquot protectors joined the long line of individuals who have practiced civil disobedience to make governments more just and democratic. This book tells some of the stories of these courageous people whose bravery captured the imagination of the world. Among the many inspiring acts of heroism were 81-year old Irene Abbey who insisted on standing trial rather than accept an offer of release; Tyson and Adam Harris, 11 and 12 years old, who persuaded their father to take them in to be arrested; two grandmothers who spent an evening in a drunk tank after police removed them from Kennedy Bridge in leg shackles; the business people who endured an attack by angry thugs while the Royal Canadian Mounted Police looked on impassively; the Anglican priest who, hearing radio reports about the Clayoquot arrests, turned his car around in Alberta and drove back to the west coast, arriving in Clayoquot after 36 hours just in time to be handcuffed with that morning's arrestees and the Anglican bishop who suddenly arrived to support him. There are hundreds more equally compelling stories in this book. With their actions, these heroes spoke for the many Canadians unable to travel to Clayoquot Sound but committed nonetheless to preserving their nation's heritage. And they spoke for future generations.

When I first visited Clayoquot Sound in the winter of 1993, I understood instantly the pride Canadians feel for the west coast of Vancouver Island, where snowcapped mountains crowd the estuaries they feed with fresh water and nutrients; where wide mudflats form one of the finest migratory staging grounds of the Pacific coast; where primordial forests of giant cedar, hemlock, and Sitka spruce are home for bear, wolf and cougar, and where great rookeries of sea lions and bald eagles congregate for the herring run.

But Clayoquot Sound's greatest asset is its people. Its greatest spectacle was the hundreds of colorful tents and shanties that crowded the flats below the giant clearcut known as the Black Hole. Its greatest drama was the thousands of Canadians and people from around the world who came to endure arrest and discomfort for the sake of future generations. Its greatest inspiration was the dissolution of ancient boundaries as the First Nations of Clayoquot Sound made partnership with local and international environmentalists to defend the age-old forests. Although the arrests have stopped for now, the power of this partnership will not subside until the clearcutting stops and native land rights are permanently ensured.

For this moment, we can reflect on the achievements of these protectors who, in an era of cynicism, alienation and disaffection, have demonstrated the force of individual acts of conscience.

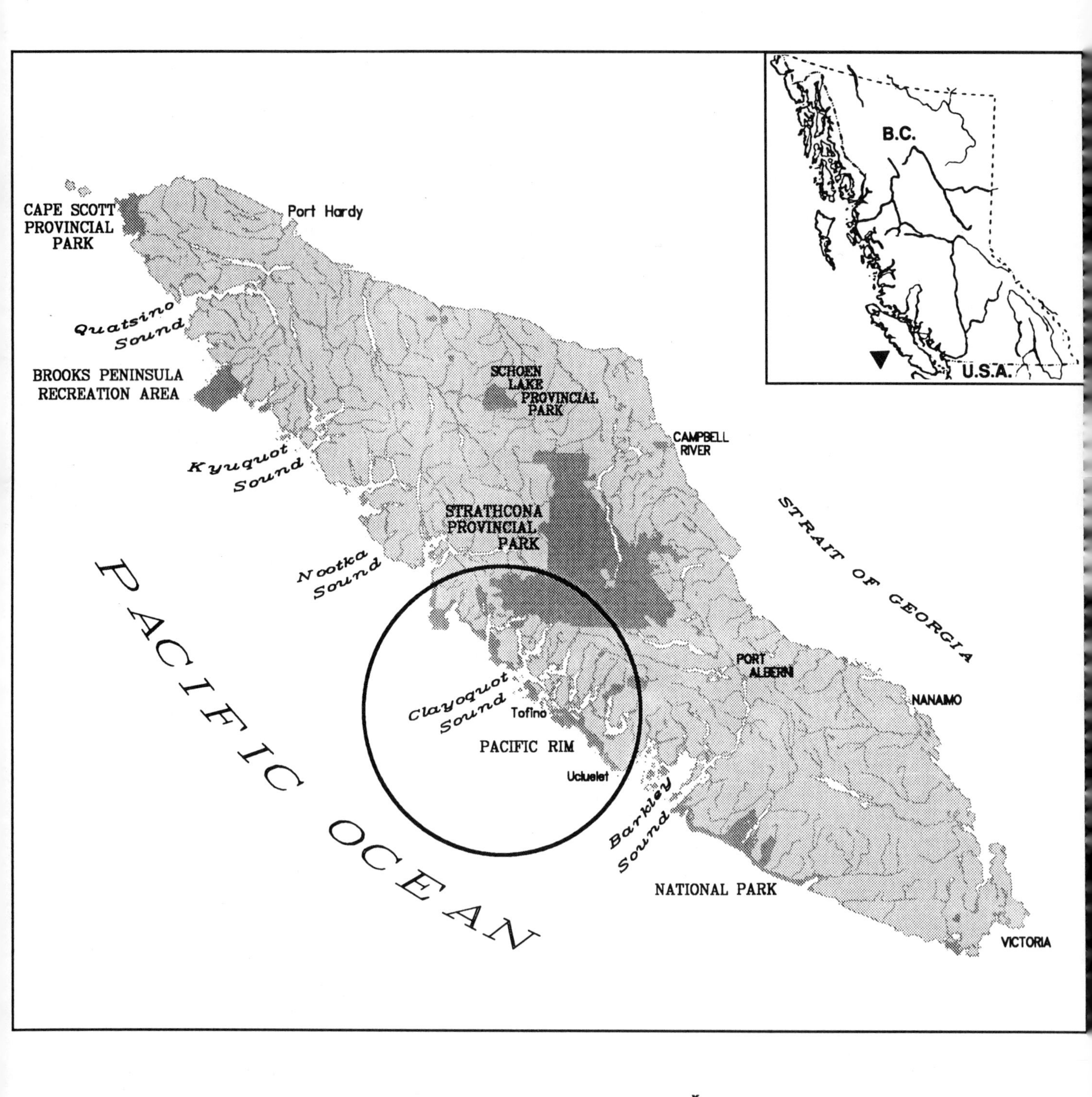
B.C.
U.S.A.
CAPE SCOTT PROVINCIAL PARK
Port Hardy
Quatsino Sound
BROOKS PENINSULA RECREATION AREA
SCHOEN LAKE PROVINCIAL PARK
CAMPBELL RIVER
Kyuquot Sound
STRATHCONA PROVINCIAL PARK
STRAIT OF GEORGIA
Nootka Sound
PACIFIC OCEAN
PORT ALBERNI
NANAIMO
Clayoquot Sound
Tofino
PACIFIC RIM
Ucluelet
Barkley Sound
NATIONAL PARK
VICTORIA

CLAYOQUOT SOUND SNAPSHOT

"It is obvious that the Clayoquot mass trials will be judged by history as one of the most important Canadian events of this century."
—John Willow Sr., Duncan, B.C.

Clayoquot (pronounced Klak-What) is a word from the Nuu-chah-nulth people whose territory includes Clayoquot Sound.

» the largest lowland temperate rainforest left in the world

» one of the largest tracts of temperate rainforest remaining on Earth

» size of Clayoquot Sound: 262,000 hectares

» location: the west coast of Vancouver Island, British Columbia, Canada

» the largest protest in Canadian history

» the largest mass trial in the western world

» the charge: criminal contempt of court for defying an injunction banning demonstrations on company work sites obtained by MacMillan Bloedel, the province's largest forest company.

» number of people arrested: 932

» range of jail terms: suspended sentence to 6 months

» range of fines: $250 to $3,000

» number of people who visited Clayoquot protection Peace Camp: over 10,000

» where they were from: across Canada, the U.S., Europe and Australia

» amount scheduled to be logged under B.C. government's April 13, 1993 land use decision: 45% clearcutting plus 17% "special logging"

» amount of old-growth to be cut: 74%

» amount to be protected: 33%—mostly bogs, marginal forest land, and alpine rock

» companies logging the Sound: MacMillan Bloedel and Interfor

» amount of public money spent by B.C. government in 1993 on a PR newspaper sent to every household in the province attempting to justify its decision to log the majority of Clayoquot Sound: $220,000

» amount of public money committed by Canadian governments to fund PR for overseas forest industry propaganda: $47.1 million

[Sources: *Canadian Forum*, July/August 1994, Friends of Clayoquot Sound; *British Columbia Environmental Report*; Western Canada Wilderness Committee Education Report, Summer/Fall 1993; *Common Ground*, July 1993.......]

PREFACE

Ron MacIsaac

My wife told me that she and her two daughters, one a psychologist and one a travel agent, were going up to Clayoquot Sound to protest the clearcutting of the rainforest by MacMillan Bloedel logging company and the provincial government's acquiescence in that destruction.

Clayoquot is a pristine area of giant trees that have taken upwards of one thousand years to grow. It is a wraparound area of the Pacific Rim National Park between the fishing villages of Tofino and Ucluelet on Canada's westernmost rim.

I set out a day ahead of them and pitched my tent in the infamous Black Hole, a clearcut on the road into the Clayoquot forest and site of the logging protestor's Peace Camp. As in the Vietnam war protests, the forest defenders were drawn from all walks of life, but on the whole were above average in education and intelligence.

Women took their share of the risks that are the hallmark of non-violent protest in the face of violence against them. Pacifism was the dominant philosophy. The special gift of women is consensus, while men are prone to confrontation; consensus was the method of choice.

At the camp I listened to the wisdom of the First Nations people who stood out against the destruction of their homeland. They spoke of the Earth and how the death of the big trees brings death to the water, the fish, the game, the healing herbs, the very regenerative tissue of the soil itself.

I found that I admired the people who stood up for the planet. They were old, young and middle-aged. They were from all over the world; I even met a dozen Basques who spoke no English, but could say in Spanish that clearcutting kills men as well as beasts.

The evening meal was out of Maxim's restaurant; the cook was from Paris. The after-dinner conversation was impressive. I went to sleep on a bed of gravel wash. In the night came a mini-hurricane that blew down the entire camp and left me floundering in floodwater. My wife and family arrived but, due to the destruction, there was no protest line to face the company bulldozers in the morning.

The next day I went at my wife's suggestion to see the start of the trials of the seven hundred people arrested at the Clayoquot bridgehead to that point. The largest trial in Canadian history had begun. At the Victoria courthouse there was chaos. Crowds of arrestees, their friends and families, clergy, media, lawyers, sheriffs, police and public filled the old building and spilled out into the street.

The first 50 were sent to the court of Mr. Justice John Bouck—in my experience a likeable, able and popular judge. Here the accused were numbered. Those who had lawyers asked for separate trials, and adjournments so that their defence counsel could clear their timetables of prior commitments.

When not enough time was granted, the lawyers withdrew save for three who could attend part-time. It was then that I, a semi-retired lawyer, decided to help with the defence. I was not in my black robes but I stepped forward and announced my commitment, little knowing that it would be the longest trial of my life—one-and-one-half months—and this only covered the first 50 forest defenders, a number that shrank to 44 by trial end.

In this book we give the reader selections from the defendants' courtroom experiences in their marathon trials. The story unfolds in the defendants' own words, and those of the press and public in response to issues raised by the trials (along with background information to set the issues in context).

Many felt the trauma of the trial was a worse punishment than the jailings that followed. The courtroom for them was a stressful steampot, a windowless room, a strait-jacket of evidence with rules that stopped their defence efforts in their tracks. The good conduct rules of these peaceful protestors were strict but some broke under the strain and there were scenes in the courtroom that upset the presiding judge. Time and again, the protestors railed at court decisions that made irrelevant their passionate evidence that their protests were necessary to save the lives of plants, animals and humans themselves.

The defenders did, however, succeed in telling their story. It was moving to hear how people from all walks of life stopped suddenly in their busy lives and offered themselves up on the altars of the Canadian justice system. The justice system remains firmly locked in the past in what the law calls precedent: basing current judgments on what the last judge did.

If we know our history, we all know that precedent brings stability to our world. We know too that many of our most cherished rights and liberties come from the activism of protest. The cutting edge of change is preferably the activity of executive branches of government or legislation. But where they fail to see peril to their citizens, one sometimes finds in the equitable powers of the judges a recognition of changing times and values and the need for immediate action. As Judge Seaton said of the destruction of forest land claimed by the First People: "The proposal is to clearcut the area. Almost nothing will be left. I cannot think of any native right that could be exercised on lands that have recently been logged."

This book is also the story of waste. One thousand-year-old trees are replaced by tiny potted plants.

The protestors in their evidence tell a vivid tale of the ongoing destruction of the Earth's last ancient forest—put to death on a pyre of greed by companies that, like ships sailing under flags of convenience, enjoy the irresponsibility of foreign and anonymous ownership.

There is a message of hope in this book. The defenders did not lay down their freedom without purpose. The world's awareness is awakening slowly and ponderously but surely. As Robbie Burns said: "It's coming yet, it's coming yet, for a' that and a' that, when men to men shall brothers be the world o'er."

It has been said that the brotherhood of man transcends the sovereignty of nations. Sisterhood and brotherhood mean more. They mean our kinship with the Earth and all things in it and on it. We are our brother's keeper. We are our sister's keeper. Here is what the Clayoquot arrestees protested for, these people who believe in the defence of the giant trees.

1

RAINFORESTS UNDER ATTACK

THE LARGE MAJORITY OF THOSE arrested at Clayoquot Sound during the summer of 1993 were moved to act by the knowledge that temperate rainforests throughout the world are being destroyed at an extremely alarming rate. They joined the blockades knowing that the effect of forest destruction on the biosphere can no longer be limited to

The Worldwatch Institute tells us that we have fewer than 10 years to turn things around or "civilization as we know it will cease to exist...."

exotic far-away regions like the Amazon. They stood their ground because, all around them, they could see the death of their own old growth forests. In one place and one time, their concern for the global and the local came powerfully together.

THE GLOBAL PICTURE

Over the past 2,000 years one third to one half of the world's forests have been liquidated.

— **Bruce Torrie (lawyer), Affidavit in the Matter of Civil Disobedience at the Clayoquot Blockade, June 7, 1994**

◆

In the past 40 years, the world population has tripled to 5.5 billion... This phenomenal growth is taking its toll as natural habitats are paved over, polluted, logged, minded and built on.

— **Val Allen, *British Columbia Environmental Report*, October, 1993**

◆

40 hectares (100 acres) of tropical rainforest [are destroyed] every minute, and with it...at least 20,000 species a year [go extinct]....

Research groups such as the Worldwatch Institute in Washington, D.C. tell us that we have fewer than 10 years to turn things around or "civilization as we know it will cease to exist...."

Rainforests...cover only six percent of Earth's land mass, but harbor more than half of all species on the planet....

Tropical forests have existed in place for more than 100 million years. Now these primeval landmarks may disappear forever within our children's lifetime.... Between 16,188,000 and 20,235,000 hectares (40 and 50 million acres) a year [of rainforest disappears], an area almost as large as West Germany....

Rainforests are, in effect, "the lungs of the planet," helping to regulate the exchange of oxygen and carbon dioxide, just as our own lungs do. In the Amazon region alone, 75 billion tons of carbon are filtered out of the air by trees....

As...trees are cut or burned, they release their carbon into the air as carb-

on dioxide, exacerbating the greenhouse effect [global warming]....

Computer simulations published in the journal Nature in 1989 raised more fears that the rapid deforestation of the Amazon would upset global and regional weather patterns. If large tracts of rainforest continue to be replaced by pasture, rainfall in the region could be dramatically reduced—the researchers predicted by as much as 20 percent.

That result would be felt worldwide. Brazil's...Environment Minister Jose Lutzenberger has gone so far as to suggest that if the forests are destroyed in the next 20 years, "the whole weather system might collapse...."

Ecologists are convinced we are the last generation that has a chance to save [rainforests]. Approximately a quarter of the biological diversity existing as of the mid-1980s will vanish over the next 25 years, as the remaining forest refuges disappear....

"All predictions are that within 50 years anything not locked up in a national park will be gone," explains Randy Hayes [of the Rainforest Action Network].

— Anita Gordon and David Suzuki, *It's a Matter of Survival,* Toronto: Stoddart, 1990.

◆

Ninety percent of their temperate rainforest...[was logged by] New Zealand. England, Scotland and Turkey logged all of their forests hundreds of years ago. The responsibility for protection of temperate rainforests now rests in the hands of the Chilean and the British Columbian governments.

— "Protect Clayoquot Sound" fact sheet, Friends of Clayoquot Sound

◆

The ancient temperate rainforests of the Pacific coast are...clustered between the Pacific Ocean and the mountain ranges that line the west coast of North America.... [They] extend from the giant redwoods of northern California to the primeval Sitka spruce rainforests of southeastern Alaska. Along the way they take in the majestic Douglas fir forests of Oregon, Washington and British Columbia, and the gnarled cedars, drooping hemlocks and stately Sitka spruce of Washington, British Columbia and Alaska....

This...ecosystem...achieved its present form 2,000 to 3,000 years ago.... The tallest trees in the west coast rainforest power their way to the sky for

Temperate rainforests are extremely rare, covering about 0.2 percent of the Earth's land area.

more than 91 metres (295 feet)—roughly the height of a 30-storey sky-scraper....

Temperate rainforests—predominantly evergreen forests that receive at least 1,000 millimetres (40 inches) of rain a year distributed over a minimum of 100 days—are extremely rare. Outside of western North America, which accounts for roughly two-thirds of the temperate rainforests in the world, they occur primarily in parts of Australia, New Zealand, Norway and Chile....

In a typical rainforest, the...trees easily last from 300 to 800 years, but some can grow to be much older....

The Douglas fir, Sitka spruce and western red cedar of Vancouver Island rank among the world's largest and longest—living plants....

— Adrian Dorst and Cameron Young, *Clayoquot: On the Wild Side*, Vancouver: Western Canada Wilderness Committee.

◆

LIFE ON THE BRINK: Temperate rainforests are probably the more endangered ecosystems. Of the 31 million hectares once found on Earth, 56 percent have been logged or cleared. In the contiguous United States, less than 10 percent of old-growth rainforests survive, scattered in small fragments throughout the Pacific Northwest. In the rainforests of British Columbia, only one of 25 large coastal watersheds has wholly escaped logging.

— Lester Brown et al., *State of the World 1992: A Worldwatch Institute Report on Progress Toward a Sustainable Society*, New York: W.W. Norton and Co., 1992.

◆

Temperate rainforests are extremely rare, covering about 0.2 percent of the Earth's land area....

Far exceeding tropical rainforests in the volume of plant biomass they contain per hectare, the temperate rainforests in North America take on their most luxuriant expression around the Olympic Peninsula of Washington State and the neighboring southwest side of Vancouver Island....

— Sierra Club of Western Canada statistics quoted in Western Canada Wilderness Committee Educational Report, Summer/Fall 1993

◆

The World Scientists' Warning to Humanity...was signed by 1,575 scientists from 69 countries, including over half [99] the living Nobel laureates.... [They] pointed out that we have at most a few decades, and possibly only one decade, in which to avert threats which will immeasurably diminish the prospects for humanity.

— Robert F. Harrington, author of *To Heal the Earth*, in *British Columbia Environmental Report*, October, 1993.

◆

WORLD SCIENTISTS' WARNING TO HUMANITY:

Forests: Tropical rainforests, as well as tropical and temperate dry forests, are being destroyed rapidly. At present rates, some critical forest types will be gone in a few years, and most of the tropical rainforest will be gone before the end of the next century. With them will go large numbers of plant and animal species.

Living Species: The irreversible loss of species, which by 2100 may reach one-third of all species now living, is especially serious. We are losing the potential they hold for providing medicinal and other benefits, and the contribution that genetic diversity of lifeforms gives to the robustness of the world's biological systems and to the astonishing beauty of the Earth itself.

Much of this damage is irreversible on a scale of centuries, or permanent.... Increasing levels of gases in the atmosphere from human activities, including carbon dioxide released from fossil fuel burning and from deforestation, may alter climate on a global scale.... The potential risks are very great.

Our massive tampering with the world's interdependent web of life—coupled with the environmental damage inflicted by deforestation, species loss and climate change—could trigger widespread adverse effects, including unpredictable collapses of critical biological systems whose interactions and dynamics we only imperfectly understand.

Uncertainty over the extent of these effects cannot excuse complacency or delay in facing the threats....

Warning: We the undersigned, senior members of the world's scientific community, hereby warn all humanity of what lies ahead. A great change

in our stewardship of the Earth and the life on it is required, if vast human misery is to be avoided and our global home on this planet is not to be irretrievably mutilated....

What We Must Do: We must bring environmentally damaging activities under control to restore and protect the integrity of the Earth's systems we depend on....

No nation can escape from injury when global biological systems are damaged. No nation can escape from conflicts over increasingly scarce resources. In addition, environmental and economic instabilities will cause mass migrations with incalculable consequences for developed and undeveloped nations alike.

Developing nations must realize that environmental damage is one of the gravest threats they face....

A new ethic is required.... We must recognize the Earth's limited capability to provide for us.... This ethic must motivate a great movement, convincing reluctant leaders and reluctant governments and reluctant peoples themselves to effect the needed changes.

— **Union of Concerned Scientists,** ***World Scientists' Warning to Humanity*****, November 18, 1992**

CANADA, BRITISH COLUMBIA AND CLAYOQUOT SOUND

CUTTING DOWN CANADA: Canada is a forested nation encompassing nearly 10 percent of the world's forests. [It has] eight major forest regions....

Nearly 50 percent of Canada's land mass is forest land, but a lush and wild image of this landscape would be a green illusion—the ancient forests are vanishing. In Eastern Canada, most of the ancient pine forests were logged off in the 1800s and replaced with early successional forests. In central Canada, the boreal forests are suffering an onslaught to feed new pulp mills. In British Columbia (B.C.), which provides nearly half of Canada's timber output, clearcuts fragment nearly every commercially viable forested valley. Most of the remaining coastal old-growth in B.C. is slated to

A great change in our stewardship of the Earth and the life on it is required, if vast human misery is to be avoided....

be logged in the next two to three decades.

Prior to 1960, before clearcutting became the dominant forestry practice, selection of the largest and best trees for logging—a practice known as high-grading—left a legacy of degraded forests. For the past 30 years the forests throughout Canada have been clearcut at an alarming rate. Despite this sad history, [the country] still contains the largest concentration of intact temperate and boreal forests on the North American continent....

Eighty percent of Canada's forests are owned and managed by the provinces. In 1990, the federal government increased its role in forest management through the appointment of a minister of forests who is in charge of Forestry Canada [now the Canadian Forest Service]. This federal agency is responsible for research, policy development, public relations and promotion of international markets. Forestry Canada is a partner with provincial forest ministries through Forest Resource Development Agreements (FRDAs), [which] have been in place since 1983.

Nearly two billion dollars ($500 million in B.C. alone) have been channelled through FRDAs to "reforest" the *not satisfactorily restocked* (NSR) lands that have not recovered from past logging or wildfires. The most recent Forestry Canada statistics reveal that despite the expanded tree-planting program, the area of land "not growing commercial species 10 years after harvesting" has continued to increase. When this area is combined with the loss from fires and insects it totals over four million hectares of new NSR land in the last ten years.

Government inventories claim that nearly half of Canada's forest land base contains mature forests. But many of these mature and old-growth forests have been seriously fragmented by a multitude of clearcuts. The truth about Canada's growing timber shortage has become apparent with mills closing and logging trucks hauling smaller logs. When viewed from an airplane, the forest landscape resembles a patchwork quilt with numerous threadbare sections.

Paradoxically, as the clearcutting increased, the employment level decreased due to mechanization that benefits both Canada's "competitiveness in the world marketplace" and private corporate profits. Canada produces fewer jobs per volume of wood cut than any other industrialized

country. Canada spends far less on research and development. There are only about one-fourth the number of foresters employed in Canada in comparison with the U.S. and Scandinavian countries.

Transnational corporations and governments have clearcut vast forested areas for short-term profit, leaving damaged and polluted water courses, lost wildlife habitat and eroded hillsides. Nearly 10 percent of Canada's once productive forest land has been so devastated that it is unable to reproduce any sort of merchantable timber at all. Aboriginal people in Canada face disruptions to their lifestyle similar to those of the Amazon Indians. Canada's boreal forest is almost equal in size to the Amazon rainforest and nearly every tree is destined to become pulp. It is no wonder that Canada is called the "Brazil of the North."

...British Columbia contains one-fifth of Canada's forested land base and nearly one-half of the country's timber volume. There is probably more temperate forest wilderness remaining in B.C. than anywhere else in North America. Within this mountainous province is a rich tapestry of diverse forest ecosystems that range from magnificent, ancient temperate rainforests to stunted high-elevation stands of spruce, to Ponderosa pine growing within dry interior grasslands.

Despite this recognized ecosystem diversity, the same logging system, clearcutting, is used (over 90 percent of the time) in every area except the drybelt country. Since 1987, logging companies and the forest service have been required to return each logged site to a free-growing state where the planted or naturally regenerated trees are taller than the competing vegetation. Up to 20 percent of each site, however, is allowed to be permanently scarred and compacted with landings and skid trails that may never grow a forest again. A 1988 government study of soil degradation concluded that the annual loss of forest productivity to the B.C. economy is approximately $80 million per year, a cost that is increasing $10 million further every year.

The vast majority of B.C.'s productive forests, including areas under First Nations land claims, were long ago signed over in tenure agreements to private industry. In 1991, the major corporations controlled nearly 85 percent of B.C.'s "annual allowable cut" (AAC)....

Half of all the timber cut in the public forests since 1911 has been cut in

At the current rate of logging, all unprotected old-growth forests on southern Vancouver Island will be logged by the year 2003.

the last 19 years. The annual cut increased to 78 million cubic metres in 1991, well above the estimated long-range sustained yield of 60 million cubic metres. In 1991, the Ministry of Forests released a report that revealed that the forest inventory was inaccurate and that the annual allowable cut was too high. Since the report was released, the AAC has been reduced in some areas and a major effort is under way to complete new analyses.

Nowhere in Canada is the battle to preserve ancient forests more intense and more crucial than on Vancouver Island, where the remaining temperate rainforest is one of the natural wonders of he world. Majestic, moss-laden, thousand-year-old Sitka spruce, western red cedar and western hemlock provide cover for a myriad of plant and animal species....

The Sierra Club of Western Canada and the Seattle-based Wilderness Society have completed the first stage of an ancient forest inventory of Vancouver Island using computerized geographic information systems and satellite imagery. The inventory found that less than half of the ancient forest that existed in 1954 is still standing today, and that only one-fourth remains in the southern half of the island. Of the 90 watersheds greater than 5,000 hectares, only five are unlogged, and only two are currently protected in a park. At the current rate of logging, all unprotected old-growth forests on southern Vancouver Island will be logged by the year 2003.

Watershed protection is a major concern of B.C. citizens.... Logging roads and clearcuts have compromised the once pristine Vancouver and Victoria watersheds. While hundreds of landslides have dumped tons of sediment...into the reservoirs from clearcuts and roads, the "experts" continue to deny that logging has damaged the watersheds. Vancouver is now considering installation of a half billion-dollar filtration system—a system whose costs are not covered by the profits from logging. Under current B.C. laws, there is no legislated protection for drinking water and no one can be held responsible if water supplies are adversely impacted.

In B.C.'s interior, watershed logging has impacted not only water supply and water quality. A few years ago, after heavy spring rains, a mudslide buried three people and a house near Kelowna. An old clearcut and skid trails were identified by the forest service as "contributing factors." In the Kootenays a similar slide covered a highway a few hours before a school

bus full of children was scheduled to pass the area....

In 1991, a new provincial government was elected with promises to double the number of parks and legislate a stronger forest practices act. [It] established a new Commission on Resources and Environment (CORE) with a mandate to design a land use strategy for B.C. and organize and facilitate negotiation processes in local and regional areas where there has been the greatest conflict. However, the government's decision [not to include Clayoquot Sound in the CORE process and] to log most of Clayoquot Sound, and its reluctance to reform the [forest] tenure system, have now jeopardized the legitimacy of CORE. Despite a flurry of government processes that seem designed to keep groups talking while the logging proceeds, land use decisions are still based on short-term economics and will continue to be made by cabinet behind closed doors. If Clayoquot Sound cannot be protected with all the public support [it has,] then there is not much hope for other forested areas....

A major impediment to reforming the tenure system is the precedent for timber companies to receive financial compensation when a portion of their tenure area is removed to create a park. When the South Moresby National Park [Reserve] was formed, Doman Industries received $30 million, despite the fact that the tree farm licence (TFL) was awarded at no cost to the original licensee. Proposed legislation may allow the establishment of new parks without major compensation costs to B.C. taxpayers. Recent amendments to the province's forest management legislation strengthen monitoring of TFL forest practices, and improve the [forest] ministry's and the public's ability to obtain information from TFL holders.

The B.C. government's pledge to double the province's parkland is now being implemented through the Protected Areas Strategy. The goal is to protect viable, representative examples of the natural diversity of the province. Although the initially proposed new park areas are predominantly "rock and ice," studies are under way to identify areas that best represent both ecosystem biodiversity and recreational opportunities. An amendment to the Forest Act now allows the government to reduce the AAC and suspend logging and road building in the [park] study areas. The question remains whether the government will respond to the conservation con-

cerns of the public, or bow to the increased pressure from industry and timber workers to continue cutting the remaining few untouched valleys.

— Jim Cooperman in *Clearcut: The Tragedy of Industrial Forestry*, ed. Bill Devall, San Francisco: Sierra Club Books and Earth Island Press, 1993.

◆

In the last few decades there have been a thousand Clayoquots, all of them roaded and raped by industry. Now we are down to this one, the only unexploited major watershed on Vancouver Island.

— Tony Eberts, *The Province*

◆

It took 100 years to log one-third of Vancouver Island's commercially valuable ancient forests. It took only 36 years to clearcut the second third.... Only nine of Vancouver Island's approximately 170 large primary and secondary watersheds (over 5,000 hectares) are unlogged. According to the federal government, at the current rate of logging, all of the Island's unprotected, commercially valuable ancient forests will be logged by the year 2020....

Our current parks, which largely consist of poor habitat lands such as mountain tops and bog forests, will not support the complete variety of wildlife populations. Our forest industry is cutting far beyond the sustainable level, even according to provincial Forest Ministry statistics....

— *A Conservation Vision for Vancouver Island*, Western Canada Wilderness Committee Educational Report, Winter 1993-4

◆

Let's celebrate our successes.... On June 22 [1994], the B.C. government decided to preserve the Upper Carmanah Valley,...Lower Tsitika,...[and] 23 [other] areas [including] the Lower Walbran....

Even with the new parks just created, less than six percent of the ancient temperate rainforest on the Island is protected.... Half of the native species now living on the Island will eventually go extinct if this is all that is ultimately left to stay wild.

— Western Canada Wilderness Committee letter to supporters, July 11, 1994

◆

If Clayoquot Sound cannot be protected with all the public support [it has,] then there is not much hope for other forested areas....

THE VANCOUVER ISLAND CORE REPORT: A RESPONSE FROM THE CONSERVATION SECTOR: On February 9, 1994, the Vancouver Island Land Use Plan was released by the Commission on Resources and Environment (CORE)....

At first glance, the 13 percent protection level might seem to be generous. But upon careful examination, we find that only 7.8 percent of the Island's ancient forests are protected. The rest is rock, ice and scrub forest types that are already well represented....

The report also falls short on protecting intact watersheds....

In the final analysis, Mr. Owen's report lays out a middle-of-the-road approach to land use planning, carefully crafted according to what he thought was politically possible.

— John Broadhead, *British Columbia Environmental Report*, March 1994

◆

WILL 12 PERCENT LEFT IN A NATURAL STATE BE ENOUGH TO PROTECT VANCOUVER ISLAND'S BIOLOGICAL HERITAGE? Herb Hammond, well-known B.C. environmental forester, said in a recent interview reported in *Earthkeeper* magazine that "about 25 to 30 percent of the [forest] land base" was needed to protect biodiversity.

— Western Canada Wilderness Committee Educational Report, *A Conservation Vision for Vancouver Island*, Winter 1993-94

◆

BRITISH COLUMBIA'S LAST STAND: British Columbia...is logging its forests at a rate of more than 600,000 acres per year, more than is cut annually from all U.S. national forests combined. Most of the trees come from virgin forests....

Because British Columbia has [virtually] no laws protecting fisheries, wildlife habitat, or endangered species, the most effective tactic for environmentalists has been to build public awareness and participation, which often shames the government into action.

— Mike Randolph, *Backpacker* magazine, September, 1993

◆

Clayoquot by the Numbers

$30.2 million Total government subsidy to logging road construction and maintenance in 1992

$411,000 1992 stumpage fees paid for timber cut in B.C.

$666,946 Cost to run Ministry of Forests in same year...

50 cubic metres Typical volume of a "large" temperate rainforest tree

1.4 million cubic metres Volume of raw logs exported from B.C. last year

6 months Length of jail sentence for Sile Simpson after her third arrest on Clayoquot Sound road blockades

6 weeks Length of jail sentence for Randy Demacedo after his second offence for mishandling waste oil transport

0 Number of jail terms handed out following discovery of logging damage to 34 of 53 Vancouver Island streams

1:3 Ratio of timber cut in 1960 to timber cut in 1993 in B.C.

20,000 Estimated job losses in B.C. as forest industry becomes "more competitive"

310 Number of job cuts planned in the Alberni/Clayoquot region by forestry giant MacMillan Bloedel by 1995, regardless of the government decision on Clayoquot Sound

987 Number of jobs cut by MacBlo in Alberni/Clayoquot region, 1991-93...

13.3 Percentage of all workers in the Alberni/Clayoquot region who are involved in logging/forestry, according to the most recent census

— James MacKinnon, August 11, 1993

CLAYOQUOT SOUND—WHAT'S IT ALL ABOUT? In 1958 MacMillan Bloedel, the largest logging company in B.C., bought the rights to log 1,000,000 acres of the west coast of Vancouver Island for $30,000. Clayoquot Sound is part of that tree farm licence (TFL 44). MB holds the cutting rights to about 70 percent of the Sound with Interfor holding approximately 20 percent, and the rest remaining as private land, native reserve and unallocated Crown land.

Weeks before the provincial government announced its April 1993 decision to allow logging in 74 percent of the Clayoquot's ancient forest, [it] invested $50 million into the company, becoming the largest shareholder.

— **Friends of Clayoquot Sound publication**

◆

THE SCIENTIFIC BASIS FOR PRESERVING CLAYOQUOT SOUND: "Extrapolation of current trends in the reduction of diversity implies a denouement for civilization within the next 100 years comparable to a nuclear winter," [says Stanford biology professor Paul Ehrlich]....

The [federal government's 1991] *State of Canada's Environment* reported that "at the current rate of logging, it is estimated that there will be no substantial ancient forest left on the B.C. coast by the year 2008...."

The remaining 700,000 hectares of unprotected ancient forest [on Vancouver Island] are being clearcut at an average rate of 20,000 hectares a year, so every tree would be gone in 35 years....

Analysis of the data tables in the recent Ministry of Forests publication *Undeveloped Watersheds on Vancouver Island* indicates that almost all of the watersheds on the West Coast of the Island have been logged....

"Nine of Clayoquot Sound's primary watersheds are essentially natural.... Six of them, the Megin, Moyeha, Watta, Sidney, Ice and Cecelia, form the largest continuous block of natural primary watersheds on Vancouver Island," [according to the B.C. government background report on the Clayoquot Sound Land Use Decision].

The government has ignored these imperatives and created 14 discontinuous "protected" areas that are widely dispersed. Despite [biologist E.O.] Wilson's warning that the "extinction rate is highest in smaller patches and it rises steeply when the area drops below 1 square kilometre

In the last few decades there have been a thousand Clayoquots, all of them roaded and raped by industry. Now we are down to this one, the only unexploited major watershed on Vancouver Island.

[100 hectares]," five of these areas are less than 100 hectares in size.... The only scientifically supportable plan is to protect all unlogged forests of the Sound.

— *The Vancouver Stump*, **Guerrilla Media**

◆

LOGGERS FACE 'NO-WIN' FIGHT: Figures supplied by the Canadian Pulp and Paper Association show that Western Europe is the [forest] industry's second most important export market after the United States, buying about 15 percent of total pulp and paper exports. Germany is the largest purchaser of pulp within Europe, followed by Italy and Britain. In terms of Canadian newsprint, Britain is the largest European purchaser, followed by Germany and then France.

MacMillan Bloedel doesn't want to [give up on Clayoquot Sound] because it fears this would encourage the environmentalists to fight over other areas slated for logging. There is a feeling in the company that after giving in on other areas such as Meares Island and the Carmanah Valley...it had to draw the line somewhere. That somewhere is Clayoquot Sound.

— **Madeleine Drohan, *Globe and Mail*, April 11, 1994**

◆

Spokesmen for forestry giant MacMillan Bloedel say the Clayoquot yield will be roughly 60 percent hemlock and balsam, 33 percent cedar, two percent fir, and five percent other species....

Much of the hemlock and cedar is pulped, on its way to becoming a variety of paper products.

In Japan, the Nippon Telegraph Company has a contract for 60,000 tonnes of Port Alberni pulp destined to become telephone books. Japanese environmentalists want the contract cancelled because the pulp may contain Clayoquot wood.

— ***The Province*, September 5, 1993**

◆

MacMillan Bloedel (MB) is the world's biggest telephone book manufacturer. So goes the old-growth forest. Yellow Pages, Pacific Bell, ITT, GTE, Nippon Telegraph, as well as German, French and Italian telephone directories are printed on Clayoquot Sound forests. They are used casually for

Weeks before the provincial government announced its April 1993 decision to allow logging in 74 percent of the Clayoquot's ancient forest, [it] invested $50 million into the company, becoming the largest shareholder.

one year and then thrown away....

— Valerie Langer, Friends of Clayoquot Sound publication

◆

ANCIENT ECOSYSTEMS WORTH PROTECTING: Clayoquot Sound has caught the public imagination around the globe....

The 260,000 hectares of densely forested islands and valleys that make up Clayoquot Sound are home to ancient Sitka spruce, western hemlock and western red cedar, many being over 1,500 years old and reaching a height of up to 300 feet.

The struggle to protect Clayoquot Sound represents a strong social movement which acknowledges the intrinsic value of life on this planet and recognizes domination over life as behavior which is no longer acceptable.

— Friends of Clayoquot Sound newsletter, Winter 1993/94

CLAYOQUOT GIANT: Canada's oldest and largest-known cedar tree towers above a canopy of hanging moss and plants so thick the top is hidden from sight.

The huge red cedar is at least 1,500 years old. It takes 17 people holding hands to encircle its gnarled base, 13.5 metres (61 feet) in circumference....

It's known now as...a special symbol at the heart of Canada's old-growth rainforest—itself unique among the forests of the world....

More than 100,000 birds use the area as either home or migratory flyway, including the [nationally threatened] marbled murrelet.

— Brian McAndrew, *Toronto Star*, July 25, 1993

◆

DRINKING WATERSHEDS UNDER THE AXE, NEW LEGISLATION UNDER REVIEW: About 300 watersheds provide the primary water supply for B.C.... Water-boil advisories [have] increased six-fold since 1986.... The B.C. Branch of the Canadian Institute of Public Health Inspectors in their submission to the Royal Commission on Health Care stated, "There are no regulations in place to ensure the provision of safe water from public water systems...." The Ministry of Forests has been the lead ministry determining

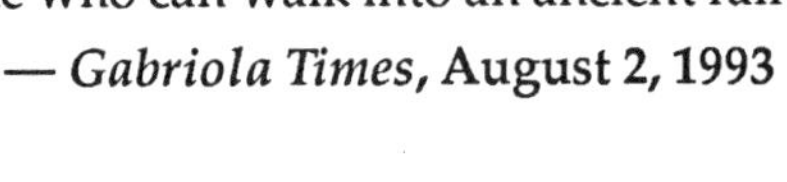

land use decisions within community watersheds....

Dr. Jerry Franklin, Chief Ecologist for Area Six of the U.S. Forest Service, has done extensive research into old-growth forests. He states that "What we find is that...groundwater coming out of the old-growth forest systems...[with its] low levels of sediment production and dissolved solids...is very high quality water."

The B.C. Forest Service has played a major role in allowing extensive clearcutting in community watersheds across B.C....

The Ministry of Forests document titled, *Ecosystems of B.C.*, and released in February 1991, states, "The quality and quantity of water within a watershed is largely a function of the intact forest cover...."

— Ken Lay, *British Columbia Environmental Report*, October, 1993

◆

It's become known as the Political Woodstock.... The trees have fallen mainly for U.S. and E.C. [European Community] timber, lavatory paper, tissues and newsprint. The government is proud: "Brazil fells 12 acres of forest a second. Well, we fell 12.9," says a trade rep....

Jerry Forrest Franklin, dean of forestry at Seattle University in the U.S. and one of the world's experts on old-growth forests...has been choppered in by MacMillan Bloedel deep into a protected area of Clayoquot Sound. He is one of President Clinton's advisers on forests:... "The value of these forests is now incontrovertible."

— John Vidal, *The Manchester Guardian*, August 21, 1993

◆

[College instructor] Vicki Levine recalls leaving for the [Clayoquot] protest with Gabriolan Kathryn Millar. "The last thing my husband said was, 'don't get arrested.' And I said, 'Don't be silly. Who, me?' "... But once she...saw what was going on..." I felt it was my body responding emotionally," says Levine.... "It unstoppered me somehow.... I don't know of anyone who can walk into an ancient rainforest and not feel spiritual."

— *Gabriola Times*, August 2, 1993

2

INDUSTRY INVADES THE FOREST

FORESTRY IN BRITISH Columbia was not always dominated by giant multinational corporations with head offices far away. Nor were forest practices as destructive as they have become today. Earlier, trees were mostly cut by small family companies, and the overall impact on the forest was minimal. Towns grew up around

sawmills, and working the forest was justifiably regarded as a way of life that might continue forever. But the mixture of corporate consolidation, new technology, and global free markets changed all that. Big business entered the picture, and trees became merely a resource to be harvested as ruthlessly as possible. The birth of the Clayoquot lease makes for shocking reading.

THE GRAB FOR CONTROL

The jury that convicted B.C.'s first Socred Forests Minister [Robert Sommers] believed he had illegally approved applications for lucrative long-term forest logging and manufacturing tenures on public lands. Specifically they believed that Forest Management Licence (FML) 22, awarded in 1955, had gone to Toronto-based financier E. P. Taylor's latest corporate venture, B.C. Forest Products Ltd., under criminal circumstances. In return they believed Sommers received a variety of bribes.... No logging companies and only two of three forestry firms implicated in Sommers' conviction were ever successfully prosecuted.... The jury could not reach an agreement on verdicts in 23 charges....

Ian Mahood, past vice-president of the B.C. Council of Forest Industries, believes that the conviction of Sommers and [timber company executive H. Wilson] Gray concealed all but the tip of an iceberg. Sommers was eventually found guilty of taking a bribe, says Mahood, but other unsavoury actions far more significant went unpunished....

Former Socred Forests Minister Robert Sommers served two years plus in jail for conspiracy. But his story is also the story of how B.C.'s provincial forests came to be concentrated in the hands of a small number of multinational logging companies....

B.C. Forest Products'...comptroller Trevor Daniels later testified that BCFP president Hector Munro told him it cost $30,000 for FML 22, part or all of it earmarked for Sommers. But as the RCMP closed in, Munro committed suicide, December 2, 1957, less than two weeks after Sommers was arrested.

FML 22 was awarded to BCFP in April 1955. But the juicy 109,000 hectare tenure was also awarded without the approval and against the advice of

Tree farm management was supposed to guarantee forests, jobs and communities forever. Instead B.C. has inherited a burgeoning leprosy of clearcuts...and tens of thousands of layoffs.

Chief Forester C.D. Orchard, in contravention of the Forest Act. Orchard had opposed it on the grounds that the application proposed a dangerous overcut. He later condemned the award of FML 22 as a timber grab and legalized liquidation, or rather liquidation with government blessing and documentary approval....

Within days of the licence award...stock skyrocketed.... $24 million clear profit [was made]....

B.C. Forest Products has disappeared into an inter-connected corporate cooperative maze that a University of Victoria researcher claimed at the time was one of four groups of companies controlling nearly the entire pubic forest land base in British Columbia....

FML 22 included the southern half of the Walbran watershed, and two blocks in Clayoquot Sound near Tofino, removed from a Public Working Circle and included in the FML amidst great public opposition....

Tree farm management was supposed to guarantee forests, jobs and communities forever. Instead B.C. has inherited a burgeoning leprosy of clearcuts...and tens of thousands of layoffs.

—**Craig Piprell,** ***Monday Magazine***

Reporter Les Leyne of the Victoria Times-Colonist reports that the lid is still on this scandal. A historian was recently refused access to the archives.

◆

HOW B.C. GOT MULTINATIONAL CLEARCUT LOGGING: The "modern era" in forest management in British Columbia began in 1947 with the provincial government's introduction of a forest tenure system featuring Tree Farm Licences (TFLs). With it came the loss of the free market system to determine the value of logs and the decline of the small logging companies, which could no longer buy rights to cut timber from the government at public auction.... The public's forest land was divided into two categories: TFL areas where large companies would have exclusive cutting rights and Public Working Circles which later were renamed Timber Supply Areas (TSAs) where small loggers were to operate on a quota system.

Only about 40 very large TFLs were awarded. Most of these went to big companies... Over the years these corporations went on to acquire more of

the cutting rights in the TSAs. This biggest government giveaway of wealth in the history of the province was not without its graft and scandal. In the case of the TFL in Clayoquot Sound, originally given to B.C. Forest Products and later sold to Fletcher Challenge and then to International Forest Products, the Forest Minister of the time, Bob Sommers, went to jail for accepting a bribe. Despite the scandal, this tenure was not revoked....

The last two forestry commissions in B.C. (Pearse, 1976 and Peel, 1991) both identified the concentration of corporate control as the major problem affecting the forest sector. But the multinational forest companies are so powerful that they have blocked all efforts to change this situation.

—*A Conservation Vision for Vancouver Island*, **Western Canada Wilderness Committee Educational Report, Winter 1993-4**

◆

The great barrier to making Vancouver Island's economy sustainable is the current forest tenure system. The various forms of tenure (government-granted rights and logging privileges) are complex and, many say, impossible to reform. They extend over huge tracts of land, 80 percent of Vancouver Island's productive forests. Their boundaries are politically, not ecologically, determined.

In B.C. over 85 percent of the wood supply is allocated to multinational logging companies that export jobs and profits along with unprocessed wood. Under this system, some predict that 2,000 jobs a year will be axed until all the old-growth is gone.

To stop this doomsday scenario from playing itself out, the big companies' tenures must be broken up. Local operators, small selective logging companies and First Nations must be given secure access to wood. Local communities reaping the profits from selection logging will reinvest them back into the local community and the local forests....

The jobs have been lost primarily through overcutting, switching to less labor-intensive and more environmentally destructive logging methods such as grapple-yarding systems, and minimal raw commodity manufacturing in mechanized mills where foreign-made computerized technologies replace workers. Only about 2 percent of the forest industry job loss in B.C. over the last 14 years can be traced to land withdrawals for new parks.

The jobs have been lost primarily through overcutting, switching to less labor-intensive and more environmentally destructive logging methods

Forestry employment on Vancouver Island, once an economic mainstay, now provides only about 7 percent of the Island's total labor force. As long as the companies continue to be subsidized by low stumpage tax breaks and bear fewer costs and receive more benefits than they should, the current job-reducing, environmentally insensitive system of forestry will continue....

Under the current industrial forestry system, set up in the 1940s over the objections of the socialists and now supported by the NDP government, the future for jobs and community stability on Vancouver Island is bleak. Timber companies and jobs now rely on huge volumes of increasingly rare and valuable old-growth wood. The second-growth forests, managed on short clearcut rotations in "tree farm" plantations, will not produce the same quality or quantity of wood.

Scientists at FORINTEK, an industry-sponsored research facility housed at the University of British Columbia, found that short rotation second-growth tree plantations that follow clearcutting are growing low value, marginally useful wood.

In 1991 Glen Manning, Chief of Economic Analysis for Forestry Canada, told the *Victoria Times-Colonist* that when the big timber is gone, small towns such as Port McNeil and Gold River will likely die. Port Alberni, he warned, will go through major shakedowns, if not death.

Forest Resources Commissioner, Sandy Peel, stated that 120,000 jobs could be lost in B.C. as the old-growth forest disappears, unless drastic changes are made to the structure of the forest industry. This means switching now to sustainable selection methods of logging in second-growth forests and carefully choosing old-growth trees for value-added products.

The parallels with the Atlantic cod industry are an urgent warning about continuing on our current course. The complete liquidation of both the ancient forests and the wild sea resources of Newfoundland has left a swath of jobless communities and ruined economies. Let's not let Vancouver Island become the Newfoundland of the West.

[Trees from] northern Vancouver Island are milled outside the region. Why? One reason is that the large forest companies are allowed to deduct transportation costs for moving timber to off-island mills from the stump-

The last two forestry commissions in B.C. both identified the concentration of corporate control as the major problem affecting the forest sector. But the multinational forest companies are so powerful that they have blocked all efforts to change this situation.

age payments they make to the government.

This is a massive subsidy amounting to millions of dollars of tax money robbed from the public purse. It is money that should go to the B.C. Forest Service to better manage the forests.

—Joe Foy, Paul George, Misty MacDuffee and Adriane Carr, Western Canadian Wilderness Committee

◆

DEBUNKING THE SIX GREAT MYTHS OF INDUSTRIAL CLEARCUT FORESTRY:

Myth: In these difficult economic times, the big forestry corporations barely break even. Their cost-cutting measures are justified even if they mean increased job loss for wood workers. If we don't do what the big companies want, they will fold their operations here and take their business and jobs elsewhere.

Reality: The major forest companies are doing very well. MacMillan Bloedel, for instance, netted profits of $1.3 billion over the last 10 years. In 1993 this company became involved in joint ventures worth $1.5 billion, all in the U.S. If big companies did vacate B.C., small business forest enterprises, which are currently limited by wood availability, could fill the gap. Right now, demand outstrips wood supply to small operators by four to one. These smaller companies currently pay more stumpage, employ more people, and add value to B.C. wood products.

Myth: Our labor costs are too high for us to compete in value-added markets. Therefore we must continue to produce in bulk raw commodities like pulp, cants, and dimensional construction lumber and leave the finer manufacturing to others.

Reality: The key consideration in value-added markets is a reliable supply of high quality raw material. We still have it....

Myth: It would cost too much to switch the industry over to high value-added production.

Reality: A recent government report estimates that by switching to speciality products, B.C. manufacturers could capture $1.2 billion in added revenue with a negligible investment [and add]...nearly 4,000 jobs.

Myth: There is nothing unusual about clearcutting. It mimics natural

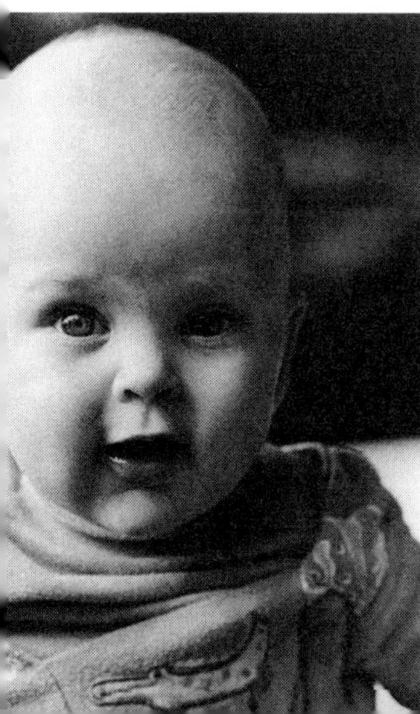

forest forces. Now that three trees are replanted on average for every one cut down, within a few years the ugly clearcuts "green up" and in 50 years no one can tell the difference.

Reality: Wild-grown, self-regenerating ancient temperate rainforests are exceedingly complex. They include snags (dead standing trees where many birds and animals live), thick mosses, lichens, insects high up in the canopy and a diversity of tree ages.... Clearcuts do not mimic fires. Human-planted forests of trees with similar genetic material allowed to grow only 80 years before they are clearcut again reduce biodiversity and in the long run are unsustainable.

Myth: Lots of small reserves will work better than large wilderness areas in protecting biodiversity.

Reality: ...Scientists believe that whole interconnected ecological units like watersheds make the best reserves.... [The] 1993 Clayoquot decision only protects three of them....

Myth: We used to build roads and log in ways that caused damage to salmon streams and unleashed mass erosion. But we don't any more.

Reality: ...To the contrary, on-the-ground improvements in logging are not yet visible.

—*Vancouver Island Paradise Magazine*

◆

Being competitive in world markets is not only a question of having an abundant supply of high quality natural resources at artificially low prices. In fact, nations like Japan and Germany with very few natural resources of their own are amongst the most competitive nations in the world. Recent information shows that Canada's competitive position, largely resulting from an abundance of high quality natural resources that Canadians did very little to create, is slipping badly.

—Ray Travers, Forester, "Forest Policy: Rhetoric and Reality" in *Touch Wood: B.C. Forests at the Crossroads* edited by Ken Drushka, Bob Nixon and Ray Travers (Madeira Park, B.C.: Harbor Publishing, 1993)

◆

Despite having the highest quality, highest volume forests in Canada, the British Columbia timber industry produces 45 percent less value added to

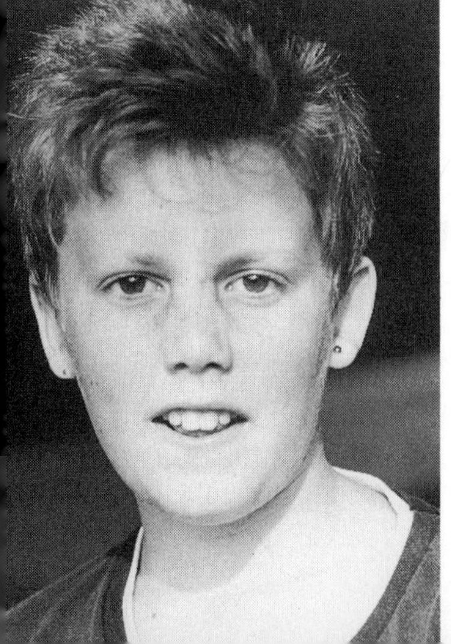

wood products than the rest of Canada.

—**Herb Hammond, RPF,** ***Seeing the Forest Among the Trees: The Case for Wholistic Forest Use,*** **Vancouver: Polestar, 1991.**

◆

Other countries—Sweden for one, Finland for another—already produce more wood from a smaller land base than we do in B.C.

—**Vaughn Palmer,** ***Vancouver Sun***

◆

TREE-CUTTING TAXES LESS THAN B.C.'S FOREST MANAGEMENT COSTS: Stumpage is a payment which logging companies make to the provincial government for the wood which they cut on Crown land. It's based on the fact that the forests of B.C. are owned by the public and the people must share the profits which the companies make through logging them. Stumpage ought to provide at the very least enough revenue to run the B.C. Forests [Ministry], which manages these forests.

Unfortunately the stumpage formula has never been fully based on the true market value of the wood, as it is in the United States where wood is valued by competitive bidding.

—**Western Canada Wilderness Committee Educational Report,** ***A Conservation Vision for Vancouver Island,*** **Winter '93-94**

◆

B.C. STUMPAGE RATES LOWEST IN THE WORLD: 1990 statistics compiled by the Food and Agriculture Organization show that Canada has the lowest overall stumpage rates in the world. As a result, other countries such as Malaysia and Indonesia use "cut and run" forest practices to minimize their costs so they can compete with Canada.

—***British Columbia Environmental Report,*** **March 1994**

◆

STREAM DAMAGE REPORT: The 'Tripp Two' report was released January 1994, based on an audit conducted by Tripp Biological Consultants of Nanaimo, B.C. The audit on compliance (more to the point: non-compliance) by logging companies to the Coastal Fisheries Forestry Guidelines (CFFG) covered 79 cutblocks on Haida Gwaii (Queen Charlotte Islands)

The major forest companies are doing very well. MacMillan Bloedel, for instance, netted profits of $1.3 billion over the last 10 years.

and on the mainland coast of B.C.

'Tripp One,' released in July 1992, exposed that 55 percent of the fisheries-sensitive streams surveyed on Vancouver Island experienced major to moderate damage from logging activities. Now, 'Tripp Two' reports that...International Forest Products multinational (Interfor) had the lowest compliance record; only 50.6 percent.... Interfor is the second largest logging corporation operating in Clayoquot Sound.

—Friends of Clayoquot Sound newsletter, Winter 1993/94

◆

AUDIT REPORT SHOWS FOREST COMPANIES CONTINUE TO DAMAGE SALMON RIVER AND STREAMS: Last year's Tripp report sent shockwaves across the province when it revealed an "appalling" picture of forest industry performance. That report found 34 of 53 surveyed streams were damaged by logging and in six cases, portions of streams suffered "complete habitat loss...."

—*British Columbia Environmental Report*, October 1993

◆

We have been slagged in *Time* magazine, *The New Yorker, National Geographic, The San Francisco Chronicle, The London Observer,* on British television, by the World Wildlife Fund and—most painfully of all—by the widely respected Worldwatch Institute, which publishes its *State of the World* reports annually.

Unfortunately...our problems are not all in the past, as working loggers will themselves often attest. Road construction is frequently lousy, landslides still occur, soil erosion is still a key Forest Ministry concern. The chief forester confirms old-growth forest inventory is being cut at unsustainable levels to compensate for the inability of second-growth stands to provide similar volumes of wood.

Native people from Clayoquot Sound are celebrating the UN's Year of Indigenous Peoples by filing complaints with the UN regarding the plunder of their traditional homeland by the talk-and-log policy of the Harcourt government.

—Stephen Hume, *Vancouver Sun*

◆

Industrial foresters have a utilitarian view of the forest. This purely utilitarian view of forests makes it easy to rationalize violence and destruction in what many people see as nature's cathedral....

Future generations will likely see the failure of industrial forestry as the most tragic misstep of all in our blind march for progress....

The industry today, like the buffalo hunter in the last century, is well aware of what is happening and will wantonly enter the last stand of majestic redwoods, of towering Douglas firs or ancient cedars and cut the oldest tree, the largest tree, the northernmost tree, the southernmost tree, even the last tree if we let it.

Loggers will do it with the same earnest dedication as the whaler going after the last blue whale.

It's what their ancestors did. It's in their blood. It is their livelihood, their lifestyle, their art.

But it is not their future. When this unsustainable orgy of cutting and running is finally over, the logger, like the cowboy, the buffalo hunter, and the whaler, will only exist in storybooks.

—David Brower, *British Columbia Environmental Report*, March 1994. David Brower is the chairman of the Earth Island Institute and a former Director of the Sierra Club.

THE INDUSTRY FIGHTS BACK

ANTI-SLAPP COALITION: Strategic Lawsuits Against Public Participation (SLAPPs) attack citizens for exercising their democratic rights. SLAPPs are lawsuits in which powerful and wealthy corporations seek civil damages for criticism expressed in a public forum. Targets are private individuals or citizen groups which disagree with the actions of the corporation....

SLAPPs depend NOT upon successful legal arguments, but rather upon exploiting sluggish judicial procedures. The SLAPP filer knows that the high costs of litigation will give a distinct advantage to the party with the greater financial ability....

The Committee for Public Participation is a broad-based coalition of groups committed to working for passage of anti-SLAPP legislation....

In autumn of 1990, the 13 largest forest companies clear-cutting B.C. hired PR giant Burson-Marsteller to advise them on long-term strategy....

[They've drafted] the Public Participation Act [and are] seeking support and endorsement for the proposed legislation.

—Vancouver Island SLAPP Defence Fund, in Friends of Clayoquot Sound publication, 1993

◆

The Attorney General Ministry is considering anti-SLAPP legislation to stop corporations from using lawsuits to stifle public debate. Ministry senior policy analyst Erin Shaw recently explained how the issue has been raised because of several SLAPPs filed against environmentalists in the province, including the now withdrawn MacBlo case against Galiano Island conservationists.... New York, California and Washington have enacted laws to protect citizens from the lawsuits.... Law professor Chris Tollefson...[who] is becoming the Canadian authority on the topic...[says SLAPP suits] are often dropped before going to court and only succeed in intimidating people into silence.

—*Vancouver Sun*, September 29, 1993, quoted in *British Columbia Environmental Report*, October 1993

◆

STRATEGIC LAWSUITS AGAINST PUBLIC PARTICIPATION (SLAPPS): Many citizens are refusing to be silenced, even when facing multimillion-dollar lawsuits. The latest trend in the SLAPP [Strategic Lawsuits Against Public Participation] phenomenon is a rise in the number of SLAPP-backs. Citizens who have been the target of a SLAPP are filing their own lawsuits, citing violations of their constitutional rights, abuse of process and malicious prosecution.

And they are winning.

—Catherine Dold, *The New Catalyst* No.25, Winter 1992/93

◆

SHARE AND THE B.C. FOREST ALLIANCE: A WHO'S-WHO GUIDE TO ENVIRONMENTAL GREENWASHING:

Share B.C.: With the growth of the popular environmental movement, corporations have helped foster counter-movements which effectively pit workers against environmentalists. In the U.S. this movement is called "wise-use," while in Canada it is named "Share." In the particular case of

Share B.C., this agency is a corporate-backed front group made up of industry workers. In the past, forest industry consultants like Patrick Armstrong have worked closely with this so-called grassroots movement providing direction and organization.

Burson-Marsteller Inc.: The world's largest PR firm and image manager [see below].

The B.C. Forest Alliance: Another "grassroots" coalition which was created by those expert architects of PR, Burson-Marsteller. In its first year, the Forest Alliance was funded to the tune of $1 million by 13 major forest corporations (including MacMillan Bloedel and Fletcher Challenge) and was led by an executive director who, at the time, was a Burson-Marsteller employee. During this initial period, the director proclaimed: "The Alliance is exploring all the issues, listening to all sides, and working toward developing a British Columbian solution to B.C.'s problems"—this coming from an employee of the world's largest PR firm, with headquarters in New York. [The B.C. Forest Alliance is not to be confused with Canada's Future Forest Alliance, a grassroots network of environmental, native, church and community groups working for greater forest protection.]

BURSON-MARSTELLER'S HALL OF SHAME: Nigeria hired B-M to put a different spin on the rampant stories of genocide during the Biafran War.

» In the late 1970s while 35,000 Argentineans were "disappeared," the ruling military junta hired B-M to "improve [its] international image" and boost investment.

» B-M "issue-managed" for corporations like Babcock & Wilcox who experienced a nuclear reactor failure at Three Mile Island; AH Robbins who felt the sting of international PR woes resulting from its Dalkon Shield IUD; and Union Carbide in the wake of the Bhopal disaster that killed over 2,000 people.

—*The Vancouver Stump*, **Guerrilla Media**

◆

PULP AND PROPAGANDA: There is a financial scandal of multibillion-dollar proportions currently unfolding in Canada's Crown forests, right alongside all the ecological devastation. From one end of the country to the

Forests, Tree Farms and Ultraviolet Radiation

In Canada the spring of 1993 saw ozone depletion of, on average, 11 to 17 percent below the 1980 average. Peak ozone depletion reached 25 percent over several Canadian monitoring stations in 1993 for periods of a week or more. A 25 percent depletion of ozone translates to a 30 to 50 percent increase in ultraviolet radiation, depending on the part of the UVR spectrum being monitored. Many studies show a significant biological impact with a 10 percent increase in UV-B radiation....

In Canada, almost all logging is by large clearcuts, where the entire forest canopy is removed. The area may then be replaced with tiny seedlings, or is left for "natural regeneration."

It now appears that UV-B radiation is a factor in seedling failure areas, particularly on south-facing slopes....

Research sponsored by the U.S. Forest Service in the late 1960s and early '70s revealed that "shading...is critical for many species, especially on south-facing slopes where heat, dryness and intense sunlight spell death for vulnerable species." This early research...is strongly suggestive that as UV-B increases under an increasingly depleted ozone layer, the negative impact on forests, especially unshaded seedlings in tree plantations, will increase.

These new developments require an immediate re-examination of the utility of clearcut logging, particularly of UV-B-sensitive species on south-facing slopes.

—Bruce Torrie, lawyer, *British Columbia Environmental Report*, October 1993

Our governments have collectively committed...$47.1 million in PR funding for overseas industry propaganda, all paid for by Canadian taxpayers.

other, the public's forests are rapidly being pulped for the benefit of a handful of multinational timber companies—a forest liquidation financed by our tax dollars....

Clayoquot Sound on Vancouver Island is a microcosm of what is happening across Canada. The two timber companies who hold cutting rights [MacBlo and Interfor] will earn millions from cutting down this ancient rainforest. Meanwhile, a recent study by Michael Mascall and Associates reveals that the province has been losing more than $1 billion per year in the forest sector. The combination of high taxpayer subsidies—in 1992 alone, $2.96 billion in federal and provincial funds to the B.C. industry—and minimal return to the government, through tax breaks and bargain-basement stumpage fees for the wood, means that the B.C. forest industry is actually a tremendous drain on the Canadian economy....

In autumn of 1990, the 13 largest forest companies clearcutting B.C. hired PR giant Burson-Marsteller to advise them on long-term strategy.... B-M has long been the PR industry's world leader in anti-environmentalist campaigns and corporate "greenwashing...."

"Given that B-M's subsidiary had already orchestrated the rise of the Share groups, sowing discord and enmity in communities across B.C., parent Burson-Marsteller advised the forest companies in 1990 to establish another organization, the B.C. Forest Alliance, which could position itself as the "rational middle-ground" on issues....

The Alliance currently claims a $2 million budget, three-quarters of which is provided by 16 forest companies. The remainder is largely provided by IWA [International Woodworkers of America] Canada and more than 170 corporate sponsors who back this "broad-based, grassroots Alliance." In 1994, Burson-Marsteller continues to advise both the Alliance and the industry.

[B-M has] long been orchestrating the anti-environmental backlash....

In early May of this year, [B.C.] Investment Minister [Glen] Clark let slip the fact that the B.C. government has promised funding to B-M's Alliance.... According to Investment Ministry spokesperson Shawn Thomas, the government grant, still in the approval state at this writing, "will allow the Alliance to set up an office in Europe for two or three years...."

The complete liquidation of both the ancient forests and the wild sea resources of Newfoundland has left a swath of jobless communities and ruined economies. Let's not let Vancouver Island become the Newfoundland of the West.

The only "misinformation" that has surfaced has been Alliance-generated.... [An] Alliance ad...run in Germany...[says] "No other forest producing region in the world has higher environmental standards for its forest practices and operations than British Columbia. Destructive clearcutting is not permitted...."

With all the talk in the province about "forest renewal" and the new forest practices code, no one's about to let slip the industry's dirty little secret: coastal rainforests don't grow back in clearcuts. Trees grow there, but they'll never be the giants that stand there now.

In a coastal rainforest, the canopy of old trees is necessary to shade the young trees as they grow. Protected from the light by the dense forest cover, the seedlings develop a "juvenile core" that grows tall and straight, and when it finally reaches the sunlight, a cellular change happens and makes the tree start to build "mature wood" in tight rings around this core....

But in a replanted clearcut on the coast, the young trees are growing in the light from day one, and so no cellular change happens. The trees remain primarily "juvenile core," with big gaps between the wobbly growth rings and wood that is crumbly and inferior, good for nothing more than pulp. Indeed, coastal second-growth lumber is so inferior that carpenters can't use it and mill foremen have to "hide" it in lumber orders, since buyers won't take it.

Obviously, the forest industry knows this. It explains...why they want to cut as much old-growth timber as they can get....

What the industry and Burson-Marsteller's B.C. Forest Alliance don't want the world to know is that coastal old-growth forests like Clayoquot Sound are not renewable resources—at least not when cut on the massive scale currently allowed. All the PR talk about "renewal" of ancient rainforests neglects to mention that the result is pulp trees that can never replace what's being cut.

Our governments have collectively committed...$47.1 million in PR funding for overseas industry propaganda, all paid for by Canadian taxpayers.

—**Joyce Nelson, *Canadian Forum*, July/August 1994**

◆

Industry Invades The Forest

For years, environmental activists have been claiming that the Share, Wise Use, People for the West and People First groups are fronts for industry, managed and funded by resource extraction interests. Now we have confirmation of these beliefs from an unexpected source: The Canadian government. Robert Skelly, Member of Parliament for [the riding that includes]...Clayoquot Sound,...commissioned a paper to be prepared by the Research Branch of the Library of Parliament ["Share Groups in British Columbia"]....

The [report's] summation reads: "With respect to B.C. Share groups, the forest companies have provided these 'local citizens' coalitions' with much of their organizational impetus and financial backing. Their apparent objective has been to pit labor against environmentalists and environmentally-oriented persons. Their effect has been to divide communities and create animosity...."

When the NDP (New Democratic Party) was elected to power in British Columbia, they began to re-examine their mandate. They had been elected by both labor (loggers) and environmentalists and they began to see that Share groups were neither. Share groups are supported by Ron Arnold and his...Centre for the Defence of Free Enterprise....

The Canadian government report is blunt. Ron Arnold is funded by the Unification Church (the Moonies) and "the American Freedom Coalition and CAUSA (Confederation of Associations for the Unification of the Societies of the Americas)." The latter two "were among the principal supporters of the Contras in Nicaragua and are backers of right-wing regimes in South America."

[American] Ron Arnold has been highly visible in Canada (fighting for pesticide use throughout the country and for forest destruction in B.C.).... Arnold's goal is to..."train interns to carry the multiple use philosophy [i.e. development even of wilderness] into every corner of Canadian society. It must initiate tactical programs of legislation, litigation and public pressure designed to change every non-timber land use designation in Canada to multiple use within 50 years."

—Naomi Rachel, in Friends of Clayoquot Sound publication, 1993

◆

Jay Mussell comes from a family of loggers, and says he sympathizes and understands the loggers' concerns about putting bread and butter on the family dinner table, and having a job....

"The corporations are not interested in trees; their interest is in money. Basically, my position on the Share Group is that they've had their share...."

—*Gabriola Times*, **August 2, 1993**

◆

Vancouver Sun reporter Mark Hume noted that at a recent seminar run by the Save Our Jobs Committee in Williams Lake, IWA Canada vice-president Harvey Arcand spoke this way about strategy:

"Environmentalists, bureaucrats, teachers or anyone else who advocates "green" positions should be harassed. Stay in their God damned face. Give 'em shit and keep 'em honest all the time and don't ever give up. That tactic was used to keep environmentalists out of Williams Lake. When the Western Canada Wilderness Committee showed up we know for instance that a motion to adjourn is always in order. We showed up and adjourned this meeting. Now there is a school of thought that says we should be cooperating with Mr. Owen; he is, after all, the commissioner.... We say, Mr. Owen is a God damned bandit. If we show up and give him any input at all it is going to give credibility to a guy who probably should not have any."

—**Mark Hume,** ***Vancouver Sun***

JOBS "VERSUS" ENVIRONMENT

Jobs? Where will you find MacMillan Bloedel when they have finished raping the Clayoquot? They will have moved on to strip-mine what is left of our forest heritage somewhere else—still huckstering jobs, jobs, jobs.

—**Charles Dobson, Sydney**

◆

As the forest companies modernize their equipment and workers are laid off, the public ends up paying for the social costs of this restructuring.... $500 million in taxpayer's money was spent to help finance the restocking of logged lands.

In 1992-93, 1.3 million cubic metres of wood were exported as raw logs from B.C.—wood that could have provided about 650 direct milling jobs. At that same time, MB invested in at least five large plants in the southern U.S.

In effect, this expenditure could be considered a retroactive subsidy.... Forest companies can construct their profit and loss scenarios so that their profits end up in subsidiaries in other countries. The value is added outside Canada....

—**Jim Cooperman,** ***Victoria Times-Colonist***

◆

I would suggest that one way out of the box that we're in would be to do everything possible to extract every job in the forest—then in the sawmill and then in the remanufacturing plant—from every stick that we harvest.... Because there is the opportunity to create probably 10,000 to 20,000 remanufacturing jobs by the year 2000.

—Corky Evans, NDP MLA, reported in *The Georgia Strait*, November 5-12, 1993

◆

LESS AND LESS EMPLOYMENT GAINED FROM MORE AND MORE TREES CUT: Forest industry giants were given exclusive "tree farm" rights to harvest B.C. timber based on the promise that they would provide jobs.

Since 1950, despite increases in the harvest levels, the ratio of people employed in the B.C. forest industry to volume of wood cut has dropped in half....

On Vancouver Island, direct forestry employment dropped by 50 percent over the last 15 years....

What has caused the loss of forestry jobs? A report by NDP MLA Corky Evans notes that it's mostly due to "increased mechanization in logging and sawmilling operations." MacMillan Bloedel refers to the changes as "restructuring." Preservation of ancient forests has not been a factor in the job losses on Vancouver Island.

DISTANT WOOD PROCESSING AND LITTLE VALUE-ADDED MANUFACTURING: The forest companies that were granted Tree Farm Licence privileges promised to build wood processing mills in B.C.... At present, 90 percent of the trees cut on the northern half of Vancouver Island is milled outside that region....

Most...B.C. wood products leave the province as large dimensional lumber, cants and minimally manufactured wood (raw pulp, wood chips and

two-by-fours). There are over forty mills...identified in Washington and Oregon which re-process B.C. wood into higher value-added products.

The continued export of raw logs from B.C. is the most blatant contradiction to the forest industry's promise of manufacturing jobs. In 1992-93, 1.3 million cubic metres of wood were exported as raw logs from B.C.—wood that could have provided about 650 direct milling jobs. At that same time, MB invested in at least five large plants in the southern U.S.

—Western Canada Wilderness Committee Educational Report, *A Conservation Vision for Vancouver Island*, Winter 1993-94

FOREST MANAGEMENT, B.C.-STYLE

When Canada became the first industrialized nation to ratify the biodiversity treaty negotiated at the Earth Summit, its well-cultivated reputation for environmental leadership got a boost....

But behind this polished veneer, B.C. has been rapidly liquidating North America's largest and most productive temperate rainforest.... Cathedral groves of ancient Sitka spruce, Douglas fir, western hemlock and red cedar are being razed at rates faster than the tropical deforestation in Indonesia and Brazil....

Today, nine multinational companies control two-thirds of the timber cut on public lands. And the archaic pricing system ensures their grip: while small timber operations must pay B.C. an average of $21 per cubic metre of wood they cut, the major companies pay just $7.

—Derek Denniston, *Vancouver Sun*, August 20, 1993, reprinted from *World Watch* magazine

◆

The government of B.C. released its new Forest Practices Code on November 9, 1993.... Clearcutting remains the status quo.... Forests have been evolving since the last ice age; reducing them to pulp plantations is hardly "high-level stewardship...."

The only penalties planned are vastly increased fines. For the Code to be truly effective, prison terms must also be specified.

—Friends of Clayoquot Sound newsletter, Winter 1993/94

◆

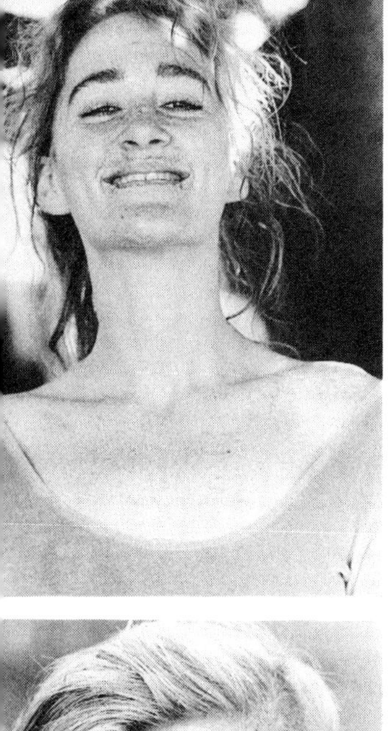

GOVERNMENT DECISION THREATENS CLAYOQUOT SOUND: On April 13, [1993] the NDP government announced its decision to allow logging to proceed in Clayoquot Sound....

The government decision claims to set aside 33 percent of the land area for park. In fact, the decision means that 74 out of every 100 old-growth trees have been or will be logged. While 33 percent of the land base may be protected under the government decision, almost half of the area set aside was already in parks (Strathcona and Pacific Rim). Of the remaining half, the protected area along the outer coast is bog, marsh and scrub trees, not ancient forests....

Few people realize that roughly 23 percent of Clayoquot Sound's ancient rainforests is already clearcut....

No wildlife studies have been conducted in Clayoquot Sound. It was the public who found Roosevelt elk in the Ursus watershed....

[Of the tourist potential:] over 500,000 people travel [the highway to Clayoquot] annually. There are opportunities for whale watching, bird watching, hiking, salmon fishing, wildlife viewing, sailing, kayaking and much more.

—Clayoquot Sound Information Sheet, Sierra Club of Western Canada, June 1993

◆

The percentages are astounding: 45 percent of the Sound is scheduled for clearcut logging, 17 percent for "special logging," with 33 percent...remaining as protected areas. MacMillan Bloedel and Interfor were given more than they had actually asked for.

—*Common Ground*, **July 1993**

◆

The Friends [of Clayoquot Sound] have obtained a B.C. Forest Service memo waiving all special management guidelines for one year in order to expedite logging in one of Clayoquot's last remaining intact rainforests.

—Friends of Clayoquot Sound publication

◆

"There isn't any possibility of MB expansion in B.C.... When we make any large investments, we'll put them where they can get the best return. That's not in B.C., and it's probably not in Canada."

SCIENTIFIC PANEL SAYS "AVOID CLEARCUTTING" IN CLAYOQUOT SOUND: Attempting to make its April '93 decision to allow logging in most of Clayoquot Sound acceptable to the public, the B.C. government set up a Scientific Panel....

The panel's March '94 interim report concludes that current forest practices lead to the loss of biodiversity. It recommends new, wholistic, "ecosystem-based" practices that:

» maintain the integrated functions of soils, freshwater, marine and forest ecosystems;
» maintain old-growth forest characteristics;
» avoid clearcutting.

The panel also recommends that road construction and logging in any of Clayoquot's pristine valleys be delayed until full resource inventories and long-term plans have been approved and exemplary forest practices have been demonstrated elsewhere.

—Western Canada Wilderness Committee Educational Report, Summer 1994

◆

Coastal temperate rainforests have always been a rare ecotype on Earth.... Today 90 percent of these wild forests are gone. The 10 percent that is left—one-quarter of it in B.C.—is disappearing at an accelerating rate....

The B.C. government claimed it could "sustain development" through its April '93 decision to allow logging in over two-thirds of Clayoquot's remaining ancient forest. Dr. Brent Ingram of the University of British Columbia's Faculty of Forestry issued a report in February of 1994 concluding that this decision is not based in science and will result in a loss of biodiversity....

The government has given...the right to cut unsustainable amounts of timber—600,000 cubic metres of wood—15,000 logging trucks worth—annually.

—Western Canada Wilderness Committee Educational Report, Summer 1994

◆

CLAYOQUOT FIRST NATIONS AND B.C. GOVERNMENT SIGN INTERIM AGREEMENT: On March 19, 1994...the First Nations of the Central Region of the Nuu-chah-nulth Tribal Council and the province of B.C. signed an historic two-year *Interim Measures Agreement.*

This legal document sets out how decisions regarding resource development in Clayoquot Sound will be made until a treaty with the First Nations, who have aboriginal title to the region, is successfully negotiated. The federal government is expected to initiate these treaty negotiations in the very near future....

The B.C. Ombudsman...confirmed that government has...failed to consult with the First Nations [about the April 13, 1993 decision on the Sound].... The interim agreement gives the First Nations virtual veto power over resource management and land use planning within Clayoquot Sound....

The First Nations are on record supporting a ban on clearcut logging the Clayoquot Sound and have expressed enthusiasm for the ecosystem-based recommendations coming out of the Scientific Panel, which was set up to provide direction for new forestry in Clayoquot Sound.

—Western Canada Wilderness Committee Educational Report, Summer 1994

◆

TAXPAYERS STUNNINGLY GENEROUS TO FOREST FIRMS: We've been hearing a lot these days about how important the forest industry is to the B.C. economy and how beholden we, the taxpayers, should feel toward it. Corporate lobbyists like the B.C. Forest Alliance and the Council of Forest Industries (COFI) have gone into hyper-gear to convince us that, without all those multinational timber giants cutting down our trees, the whole provincial economy would simply go belly-up.

In fact, the reality is quite the reverse. It is the B.C. taxpayers who have been incredibly generous to the big corporate loggers. For years, our provincial government has been giving the multinational companies stunning financial concessions, including bargain-basement stumpage rates for old-growth timber. Expert consultants estimate that these stumpage rates range from "one-quarter to one-eighth the true market value" of the re-

As both jobs and trees continue to be clearcut, the timber barons go on laughing all the way to the (U.S., Japanese, and New Zealand) banks.

source.

The result is revealed in the first column of the accompanying chart. B.C. Ministry of Forests (MoF) *Annual Reports* indicate that for six out of the last ten years (1983-1992), the Ministry has been operating in the red, including a net loss in 1992 of $59 million. Over this ten-year period, the B.C. Ministry of Forests has registered *total net losses* of $711.8 million. During the same period, the volume of timber cut in B.C. climbed at an unprecedented rate.

... The public has a right to know the reasons for its increased tax burden. During the same period in which the B.C. MoF has been losing so heavily, our personal income taxes increased at an annual average rate three times faster than corporate taxes. According to a *Times-Colonist* editorial (September 30), by 1991 in Canada, "individuals were paying 90 percent of all income taxes, while the corporate contribution had shrunk to 10 percent."

In B.C., much of that discrepancy is the result of years of "sympathetic administration" of the forest industry. Our governments have allowed the

THE BALANCE SHEET
All figures in millions of dollars

	BC MoF Net Revenues	MB Operating Earnings	MB Deferred Canadian Income Taxes (recovered)	R&D Gov't Grants to MB	MB Investment Tax Credits
1983	-217.9	40.5	13.1	0.6	NA
1984	-137.1	67.2	12.4	0.6	NA
1985	-137.3	108.9	18.8	1.2	NA
1986	-234.3	222.5	26.1	1.9	5.5
1987	-192.1	465.8	123.7	1.3	4.1
1988	1.4	488.0	29.7	1.2	3.8
1989	142.2	365.1	29.6	0.8	3.9
1990	113.2	91.3	(20.4)	0.8	3.2
1991	8.7	-117.5	(42.3)	2.4	3.3
1992	-59.0	66.8	(9.1)	3.1	2.7
	-711.8	**1,798.6**	**253.4 ($71.8)**	**13.9**	**26.5**

big corporate loggers to defer hundreds of millions of dollars in taxes owed—a loophole not available to the average citizen. As well, we've been picking up much of the tab for all the tree-planting needed to be done in the wake of their grapple-yarders and feller-bunchers. In 1983, the Forest Resource Development Agreement (FRDA I & II) committed $800 million in federal and provincial tax dollars for replanting in B.C.'s overcut forests.

Moreover, as the industry "downsized" over the last ten years—replacing thousands of workers with high-tech machinery—it left in its wake an incredible swath of social destruction. According to Statistics Canada, between 1981 and 1991 at least 27,000 direct forest industry jobs were lost in this province. The Forest Resource Commission (FRC) has accurately called the multinationals-based industry a "major dis-employer in B.C."

What this means is that we, the taxpayers, have been stuck with an incredible Unemployment Insurance (UI) bill to pay—at the same time that we've been losing hundreds of millions of dollars on the cutting of our Crown forests. To get a sense of the gargantuan proportions of this social assistance bill left in the taxpayers' lap, we need only consider one town (Port Alberni) in one year alone (1990): UI payments for 3,536 claimants totalled $27 million. Meanwhile, as both jobs and trees continue to be clear-cut, the timber barons go on laughing all the way to the (U.S., Japanese, and New Zealand) banks.

We taxpayers even pay for their industry propaganda. As forest historian Ken Drushka has revealed, for many years and until very recently almost half ($6 million) of COFI's $14 million annual budget came out of federal and provincial coffers—including more than $2 million per year from the B.C. public purse. As well, *Maclean's* (August 16) informs us that another industry lobby, the Canadian Pulp and Paper Association, has been promised $4.5 million in federal and provincial tax dollars to help run its new office in Brussels for "protecting foreign markets and countering environmentalists' claims."

Indeed the multinational timber giants have been so well rewarded by governmental largesse that in 1991, the Forest Resources Commission observed (no doubt after scrutinizing the years of B.C. MoF losses, industry tax deferrals, and corporate profits leaving the province): "A 10 percent

reduction in logging and forestry activity would produce a net effect of less than two percent on the provincial GDP (Gross Domestic Product)."

Obviously, the forestry multinationals have a vested interested in disguising the taxpayer burden and clouding our perceptions. This mystification has been especially obvious in the debate about Clayoquot Sound. We're being told to think that by allowing multinational MacMillan Bloedel (MB) to continue clearcutting in Tree Farm Licence #44 in the Alberni-Clayoquot region, somehow an economic benefit will trickle down to the B.C. taxpayers and jobs will be saved. Indeed the Harcourt government's decision to purchase $50 million worth of MB shares may have been based on such a hope.

But as recent events indicate, there is no realistic basis for such a plan. MB's latest announcement (September 30) of another mill closure in the region brings the number of direct forest jobs lost there over the past four years to nearly 1,500—or approximately 4,000 over the last 15 years.

In fact, recent documentation tabled with CORE (the Commission on Resources and Environment) reveals that over those years, the "social-economic profile" of Vancouver Island has changed considerably—partly in response to its major dis-employer, MacMillan Bloedel. The study indicates that, according to 1991 figures, only 7.3 percent of the Island's total labor force is directly employed by the forestry sector. If industry layoffs during 1992 and 1993 were factored in, the figure would be even lower.

Even Port Alberni has recognized the reality of the 1990s. According to the *Alberni Valley Times* (September 29), the Official Community Plan compiled by the city's Advisory Planning Commission states that the tourism industry "will continue to be the major focus for economic development in the community."

Thus, the Harcourt government appears to be clinging to a highly outdated and invalid perception of this province (and especially Vancouver Island), as economically dependant on the forest multinationals. Even its own figures (cited in the trendy looking Clayoquot Sound mailout) pale by comparison to the real taxpayer burden resulting from the ways this industry has been administered.

The government mailout states: "In 1991, forestry contributed about

MacMillan Bloedel, having reaped incredible profits from B.C.'s Crown forests, is aggressively expanding south of the border and is not at all interested in either a sustainable forest industry or a sustainable economy in this province.

$136 million to the Clayoquot-Alberni region's economy and $23.5 million to the provincial treasury." What it doesn't tell us is that MacMillan Bloedel's style of forestry contributed more than 1,000 layoffs to the region that year, driving the region's UI payments up to the $30 million mark. And it doesn't tell us that the Ministry of Forests netted a mere $8.7 million for all the clearcutting done across the province in 1991.

For the sake of its (now outdated) figures, the Harcourt government is willing to imperil the livelihoods of tens of thousands of people on Vancouver Island, including the tourism goals of Port Alberni.

Meanwhile, MacMillan Bloedel is certainly in no danger of going under economically, even without Clayoquot Sound. Column 2 in the chart indicates that the last ten years have been good for MB, despite the whining in their recent *Annual Reports*. They've reaped $1.79 billion in operating earnings, or (if you prefer) $1.03 billion in net profits. As well, our governments have generously given the company $13.9 million in R & D grants, and $26.5 million in investment tax credits—a total of $15.5 million in the last three years alone. But that's just the beginning.

Column 3 in the chart indicates that, from 1983 through 1988, MB deferred $253.4 million in Canadian federal and provincial income taxes. Revenue Canada cannot collect deferred corporate taxes, nor can it charge interest on outstanding amounts. During the years 1990-1992, when MB's profits fell below the standard annual bonanza, the company was then able to recover $71.8 million in deferred Canadian tax.

MB's *Annual Statutory Reports* reveal other examples of government largesse (and taxpayer woe), as in this passage from 1935.

> "In response to submissions to the B.C. Minister of Finance, a number of inequitable taxation policies were changed, not the least of which was the phasing out of the property tax on machinery and equipment. When they are fully eliminated in 1988, the pretax annual savings are estimated to amount to $25 million for the Company."

Since 1988 MB has thereby saved yet another $150 million in its tax statements. Nevertheless, MB started crying "hard times" in the 1990s and the posture worked—especially in 1991. In addition to the $2.4 million in R & D grants, the $3.3 million in investment tax credits, and the $42.3 million in

recovered Canadian tax deferments, the Company's 1991 *Statutory Report* states:

> "In late 1991 the Company entered into a five-year $45 million agreement with the Canadian federal government to assist the Company in continuing, during the current difficult economic times, its wide variety of research and development programs in environmental protection, computer control and wood-based building materials development."

Through this 1991 Memorandum of Understanding (MOU), the Canadian federal government agreed to contribute $15 million to MB over the following 5-year period.

Obviously somebody should have sent the Mulroney feds a copy of the July 1991 issue of *Canadian Business*, in which MB President Robert Findlay bluntly stated: "There isn't any possibility of MB expansion in B.C.... When we make any large investments, we'll put them where they can get the best return. That's not in B.C., and it's probably not in Canada."

Whether our governments were noticing or not (it was, after all, a provincial election year), 1991 saw an incredible round of MB layoffs in B.C. By mid-summer, Monty Mearnes (second vice-president of IWA Local 1-85) was telling the press: "How many job losses do we experience before we challenge MacMillan Bloedel's right to tenure? IWA-C1-85 go on public record demanding that TFL 44 be removed from MacMillan Bloedel."

But our governments still didn't get the hint. Indeed about the same time that the MacBlo/Mulroney suits were conferring on that $45 million MOU, the company's crews were busy dismantling the Alberni Plywood mill and shipping $200,000 worth of machinery down to MB's Pine Hill, Alabama plywood facility.

While bemoaning "the current difficult economic times," in 1991 MB formed a joint-venture with Trus Joist International (TJI), a Boise, Idaho company considered the largest engineered wood manufacturer in the world. According to TJI president Walt Minnick, MB "wanted to buy TJI outright," but agreed to form Trus Joist MacMillan (TJM); with TJI putting in 11 of its 14 U.S. plants, and MB picking up the costs for three start-up plants in the U.S.

"IWA C1-85 go on public record demanding that TFL 44 be removed from MacMillan Bloedel."

By late September of 1993, MB's board had approved a more than $200 million capital-spending plan for TJM, including a new $130 million plant in Hazard, Kentucky and another $100 million to expand the three existing plants and start another. As well, the city of West Sacramento, California, has given "conditional approval" to a $1.5 billion joint-venture proposal between MB and Haindl Papier of Germany to build a recycled newsprint facility there.

It has been obvious for years that MacMillan Bloedel, having reaped incredible profits from B.C.'s Crown forests, is aggressively expanding south of the border and is not at all interested in either a sustainable forest industry or a sustainable economy in this province. Nevertheless, our government still hasn't figured this out.

On June 1, the Harcourt government released another document relating to the Clayoquot Land Use decision. In the appendix we find that since 1990, $7.2 million in federal tax dollars has been poured into Ucluelet and Port Alberni for necessary job-retraining and "human resources planning"—including (since May of 1991) $2.72 million in federal and provincial Industrial Adjustment Service (IAS) money for MacMillan Bloedel/IWA Local 1-85 joint projects. MacMillan Bloedel has put up IAS "matching grants" amounting to $425,000.

After decades of generosity from the B.C. taxpayers to the industry in general and MacMillan Bloedel in particular, the company now has the opportunity to make a significant gesture of goodwill.

MacMillan Bloedel can request that the B.C. Supreme Court drop all charges against the Clayoquot Sound arrestees, and it can gracefully relinquish tenure in TFL 44: donating back the remaining forests to the people of this province whose tax dollars have served the company so well.

—Joyce Nelson, *Victoria Times-Colonist*, October 19, 1993. (Vancouver Island writer Joyce Nelson is the author of six books including *Sultans of Sleaze: Public Relations and the Media*. Recently, two of her articles were listed in Project Censored Canada's Top Ten for 1993.)

NO TREES
NO FISH
NO JOBS

❸

PROTECTING AND BLOCKADING

THE LIQUIDATION OF THE ancient forests has been going on for a long time. In watershed after watershed, ordinary people have protected forest ecosystems, fighting battles in their defence. But, as yet, the struggle has not been won. The massive destruction has been countered by a stubborness of defence which may yet make the destroyers

Some 100 groups of all sizes, from Greenpeace International to university groups, have begun Clayoquot campaigns around the world and the campaign is growing.

yield to public opinion. That opinion is strong, diverse, and increasingly backed by moral authority.

PROTEST FROM EVERY QUARTER

CLAYOQUOT SOUND—WHAT'S IT ALL ABOUT? Over the past 15 years in Clayoquot Sound, environmentalists have participated in five different government committees and commissions dealing with forest issues. Clearcut logging has continued unabated during these processes. Over the same period of time there have been five blockades [and] over 932 arrests.

—Friends of Clayoquot Sound publication, 1993

◆

500 teenagers from good Victoria homes protest on the lawn of the provincial legislature against the logging of Clayoquot Sound. The new face of the Clayoquot protest is the face of a generation that is young and unlined and naive and angry.

—Robert Sheppard, *Globe and Mail*

◆

Canadian writers and artists support the movement to stop clearcutting in Clayoquot Sound. Al Purdy, William Gibson, Joy Kogawa, Lillian Allen, Audrey Thomas [and] Pierre Berton [will do a] benefit concert in Toronto. A transcontinental train will bring people from across the country for the reading.

Earlier on the steps of the legislature a dozen writers attracted an audience of about 350 people.

William Deverell organized the reading and auction along with writer Brian Brett. Items offered included George Bowering's baseball glove, French garters from Susan Musgrave, Farley Mowat's briar pipe, a pastel by Bill Reid, the use of Deverell's house in Costa Rica for a week, books by Jane Rule and P.K. Page.

—Chris Dafoe, *Western Diary*

◆

[Questions raised internationally include:] Canada's compliance with our

international treaty obligations—especially the UN Conventions on Biological Diversity and Climate Control.

—Chris Hatch, Friends of Clayoquot Sound newsletter, Winter 1993/94

◆

On July 1st, [1993] groups in Japan, Germany, England, Sweden, Australia, Italy and India protested the logging of Clayoquot Sound outside of Canadian embassies and consulates.... Some 100 groups of all sizes, from Greenpeace International to university groups, have begun Clayoquot campaigns around the world and the campaign is growing.

—Valerie Langer, Friends of Clayoquot Sound, in *British Columbia Environmental Report*, October 1993

◆

CONGRESSMEN WADE INTO CLAYOQUOT DEBATE: A group of influential politicians is pressing U.S. Vice-President Al Gore to take action to protect Clayoquot Sound.

Congressmen John Kerry, John Porter and Henry Waxman sent a letter to Gore urging him to press Premier Harcourt to preserve Clayoquot Sound....

"Unfortunately, our country's demand for timber is fuelling the destruction of the ancient forests of British Columbia and is undeniably a factor in the decision to log Clayoquot Sound," wrote the congressmen.

"It is unconscionable to allow U.S. demand for timber to destroy the remaining temperate rainforests of North America simply because they are outside our borders," they wrote....

Tamara Stark, spokeswoman for Greenpeace, said 25 organization members from Canada, the U.S., the U.K., Germany, Holland, Austria and France [staged] demonstrations.

—Richard Watts, *Victoria Times-Colonist*, November 9, 1993

◆

A California senator and a B.C. actress included a pledge to preserve old-growth forest in British Columbia in marriage vows exchanged in Clayoquot Sound.

Tom Hayden and Barbara Williams were married Saturday by a Buddhist priest from New Mexico and blessed by a native artist....

"Marriage is similar to old-growth forest—you have to be especially loving and vigilant to help it survive and grow."

—*Victoria Times-Colonist*, **August 10, 1993**

◆

KENNEDY, NATIVES TO JOIN HANDS: Clayoquot natives and Robert Kennedy Jr. have formed an alliance that may lead to natives joining the Clayoquot Sound blockade....

Kennedy, an American environmental lawyer [from the Natural Resources Defence Council, Washington, D.C.], and Chief Francis Frank of the Tla-o-qui-aht First Nation said they'll work together to fight for native rights in Clayoquot Sound.

—Charlie Anderson, *The Province*, August 1, 1993

◆

"How many committed environmentalists understand what has been done and what continues to be done to First Nations people? How many of us have a good class analysis of the politics of power and greed? We have to talk, person-to-person, with the First Nations people.... Talk to the Elders.

—Sile Simpson (arrestee), Friends of Clayoquot Sound newsletter, Winter 1993/94

◆

On January 24, 1994 the World Conservation Union (IUCN) at its 19th Session in Buenos Aires passed a resolution to protect B.C. rainforests.

—Ecological Rights Association, *British Columbia Environmental Report*, March 1994

◆

JIM WILLER'S STATEMENT TO THE COURT: As an artist for 41 years, I have spent my life...celebrating nature in my work. That's why I had to go to the Kennedy Bridge....

It seems that there is a blindness in us that prevents us from working in harmony and having respect for nature....

I believe that we who stood there...are presenting *an injunction* on behalf

Over the past 15 years in Clayoquot Sound, environmentalists have participated in five different government committees and commissions dealing with forest issues. Clearcut logging has continued unabated during these processes.

of nature to the government and to the court....

Let the Raging Grannies sing..."Rage, rage against the dying of the trees."

—**Jim Willer, Gibsons, B.C.**

◆

Spiked trees have been found in the environmentally disputed Clayoquot Sound area near Tofino. "It is a conscious and deliberate attempt to inflict injury," IFP [International Forest Products] operations manager, Dean Wanless, said. The tops of the spikes were cut off to make them more difficult to see, he said.

Tree-spiking has been advocated by some environmental groups as a way to keep people from cutting old-growth forests. The spikes shatter and send pieces of shrapnel flying when hit by a chainsaw. Val Langer, a Friends of Clayoquot Sound director, said her group is not responsible. "Our organization has a very clear policy of nonviolence, which includes no damage to property," Langer said.

She said she found it ironic that a potentially damaging incident for the environmental movement was made public shortly after a forest company has been through an embarrassing episode.

MacMillan Bloedel was recently fined for cutting an area where logging was restricted, she said. "It happens every time," Langer said. Friends of Clayoquot are not suspected in the recent tree-spiking, Wanless said. "We are not accusing them," he said. "They promote nonviolent direct action."

—**Dirk Meissner**

CD: THE ONLY OPTION LEFT

Having exhausted all legitimate avenues for influencing the government's decision to continue with the clearcutting of Clayoquot Sound, protectors were left with few options but to place themselves, literally, on the line in front of the logging trucks. In this, they were pursuing an honorable tradition of nonviolent civil disobedience.

Mahatma Gandhi taught that disobedience should persist in spite of everything that those in authority could do.

Mere passive resistance later evolved into the revolutionary concept of

satyragraha. Satyragraha is not mere civil disobedience; its objective is to transform relationships between people in a way that not only effects a change in policy but restructures the situation that led to the conflict. Attitudes and people are transformed.

This discovery was a weapon with which the weak could fight the strong. It may be the greatest discovery of our century. Its essence is massive, nonviolent action against tyranny and injustice; this shakes unjust authority and redeems tyrant and victim alike.

—Clayton Ruby, lawyer, *Globe and Mail*, October 26, 1993

◆

Susan Watson, a Catholic lay missionary from Toronto, [said]: "As a Christian, I practice a religion whose central figure is a man who was arrested and executed for peacefully and nonviolently struggling against the unjust economic and social structures of his time. Ancient and sacred are these forests and I stood to protect them as an Italian would protect the Sistine Chapel or a Greek the Parthenon."

—Justine Hunter, *Vancouver Sun*

◆

Tawnee Chipps, a 10-year-old boy who was determined to stay on the road despite the apparent reluctance of police to take him in to custody, said he was not going to stand by while our rainforest is being cut down.

—*Globe and Mail*, August 10, 1993

◆

81-year-old Summer Pemberton, who describes herself as a peace and environmentalist activist, is the oldest person to be arrested so far.

—Stephen Hume, *Vancouver Sun*

◆

Peace camp head cook, Guido Kettler... has cooked in Switzerland, Germany and Colorado but said the Clearcut Cafe has been his most rewarding work experience.... Kettler arrived at the Peace Camp in July with the intention of staying for one day.

—Ray Smith, *Victoria Times-Colonist*

◆

The Black Hole clearcut is an area logged twenty years ago and replanted unsuccessfully four times.

SVEND ROBINSON AND THE HALF-WITTED STATE: Having pleaded guilty of criminal contempt of court, Svend Robinson yesterday was sent to jail for 14 days....

Mr. Robinson, NDP member of parliament for the Vancouver suburban riding of Burnaby-Kingsway, violated a B.C. Supreme Court injunction last year when he and others blocked a Clayoquot Sound logging road....

On the radical left, civil disobedience is condemned as an act of resistance because it accepts the existing political structure—which is true: civil disobedience is a symbolic or ritualistic violation of law, rather than a rejection of the system as a whole. But on the conservative right, it is argued that the logical extension of civil disobedience is anarchy, and the right of the individual to break any law at any time....

It is a moral act....

No one human being...can pronounce what shall be moral or immoral for other human beings....

Civil disobedience is...a personal recognition that one is obligated by a higher, extra-legal principle to break some specific law.

—Michael Valpy, *Globe and Mail*, July 27, 1994

◆

Svend Robinson...gave instructions to the Commons administration to dock him pay for the time he's incarcerated.

—Glenn Bohn, *Vancouver Sun*

◆

It appeared to me that the women, with their nurturing attitude to the Earth, were the initiators of the defence of the forest ecosystem at Clayoquot.

Working with groups on women's [issues], very often it would seem to me to be the same problems in a different costume. Always it was the problem of privilege, of power and money, of patriarchy, of hierarchy, of laws created by and for the powerful. The only weapon people have in these situations is one of civil disobedience, when all other peaceful means fail.

—Betty Krawczyk, arrestee

◆

Satyragraha may be the greatest discovery of our century. Its essence is massive, nonviolent action against tyranny and injustice; this shakes unjust authority and redeems tyrant and victim alike.

At the...Peace Camp,...they all agree:...male dominated patriarchal society has brought the planet to the brink of environmental disaster, and it is women who will lead the way to a better world....

The prominent role women are playing in the campaign to stop clear-cutting in the Clayoquot has been mostly overlooked in media reports. And yet it is women who are the key organizers.

While it is men who run the vegetarian kitchen at the protest camp, women form the majority of people arrested and charged at the blockades.

There have been two "women and children's days," when only women and their children blocked the MacMillan Bloedel logging trucks....

Running an ecofeminist protest campaign means there is no tolerance for Paul Watson-type macho environmentalism.

—**Stewart Bell,** *Vancouver Sun*

◆

THE WOMEN OF CLAYOQUOT: The summer of 1993 began with the arrest of the First Fifty. Then in daily headlines throughout the summer, the number of arrestees mounted, and is mounting still. One day it was women and children at the blockade, then seniors, then people from various Gulf Islands, then "people of faith." A busload of business and professional people headed for Clayoquot. A busload of artists, a busload of writers.

Gradually another story began to be understood. The story of a "Peace Camp" in an ugly clearcut, run, it was said, on "ecofeminist principles." A society with a peace code: openness, friendliness and respect for all living things, no verbal or physical violence or damage to property, no weapons, alcohol or drugs, an atmosphere of calm and dignity. An ad hoc teaching institute under plastic sheeting and rough wood, actively promoting three cardinal rules: equality, nonviolence and decisions by consensus.

"I wish they would keep this camp going," said a tanned women from Saltspring on the last day of the camp. "So many teenagers wanted to come and have not made it yet. And it's such good training!" She held her hand to her heart and smiled. "My mother's heart."

On the last night of the camp the young male cook spoke seriously, "This has been a transmutative practice for me. It has been beautiful to see people of such wakefulness in a time when we are surrounded by death." Sur-

rounded by death they were. The Black Hole clearcut is an area logged twenty years ago and replanted unsuccessfully four times. Blackened stumps protrude from slash left behind. The soil has eroded to sharp stones, criss-crossed by rusted tree-hauling cables.... There were stories of families estranged, a car trashed, a bus hijacked in the middle of the night. The camp itself was under seige, with persons arriving every day at the gate, sometimes drunk and confrontational, leaning on the horns of their vehicles in the middle of the night, even managing to creep in across the clearcut to throw stones at the tents. Nasty encounters seemed written into the scenario. Thousands of people were passing through the camp, ten thousand by summer's end. How could this transient and often angry population be brought within the idealistic principles of the camp's origin?

It happened because, first of all, it was carefully planned. Tzeporah Berman, 24, a graduate student in environmental studies at York University, helped choose the sites for the camp and the blockade. "I was very concerned about making sure that the protest would be peaceful. We organized training in nonviolence every day. We developed the circles as a way of decision-making, and worked on how they could be facilitated, and how to make people feel a sense of ownership." Peacekeepers were trained. Relations with the RCMP, at first poor, became friendly.

Jean McLaren, 66, from Gabriola Island, offered to come for the month of July. "Jeannie," as she was called in the camp, has facilitated workshops in nonviolence at the Nevada nuclear test site and for two summers on the Israel Peace Walk. She stayed until the camp ended in October.

Dana Kagis, 19, went to the Clayoquot area to visit her mother and to look for work in Tofino. She thought she would visit the blockade once in a while. Instead, like Jean, she stayed all summer, helping to facilitate the workshops. "When the Midnight Oil concert happened, I got very little sleep for three days. I was running on pure panic. We handled it. We don't know how! We had to form another camp down by the bridge for four thousand people. We tried to give the nonviolence workshop to as many people as possible and I think we hit most of them."

Dana laughs as she remembers one of her favorite stories from the summer. "There were four or five men near to the front when the concert was

going on, with a sign saying 'Go Home Aussies to Your Burning Beds.' They were holding up their sign so that people couldn't see. Jean went over to them and talked with them and made them promise that they would behave. And every time they started not to, she would go, 'Oh! You promised!' and they would lower their sign."

"It was often a struggle to keep the idea of Ghandian-style nonviolence as our focus," admits Valerie Langer, 30, "but we did it. Sometimes there were over 400 people coming to consensus over the nature of the action we would take. I have no doubts the Peace Camp changed the lives of hundreds, thousands of people." She quotes one retired labor organizer who spent three days there: "This shouldn't work, but it does." Valerie worked both in the camp and in the office of the Tofino-based Friends of Clayoquot, working on the international campaign to save the remnants of the old-growth forest.

"We based our actions on the Women's Pentagon Action and the Women of Greenham Common," says Tzeporah Berman. "Certainly this camp was not just run by women. But women are in a special position working on environmental issues. Against them are male loggers, mainly male RCMP, a male prosecuting judge; I don't think that's a coincidence. We're not just fighting for this issue but for recognition and equality."

She now faces a virtually unprecedented court action: *five hundred charges* for aiding and abetting civil disobedience. Her trial is slated for December 6. "If they win this, it sets them up to arrest other people who are facilitating social justice and environmental actions and to intimidate others from organizing."

Most of the camp organizers face stiff jail sentences and large fines. Sile (pronounced "Sheila") Simpson, 42, has the distinction of having received the longest, most punitive sentence so far: she has been arrested three times for blocking logging roads in the Clayoquot. On the day of her third arrest this past July, she had not planned to be arrested again. She had taken guests from her Tofino bed-and-breakfast to see the ancient trees on Meares Island [and] took them to see the blockade. Suddenly she found herself once more blocking a large logging truck. She was given a six-month prison sentence, later reduced to four months on appeal, was imprisoned

at Burnaby Correctional Centre for Women for twenty-two days and then released on electronic surveillance.

Later in the summer, she sat at the gate of the Peace Camp singing songs she has composed about the blockade and accompanying herself on a folk guitar.

"We women need to connect with our blood," she said in her soft Irish accent. "I am trying to get in touch with the wilder, freer spirit inside myself."

Part of the joy of their leadership for all of these women has been watching others, especially young women, come into the camp quiet and, as Langer says, "not saying boo," and then develop into positions of strong leadership. A 17-year-old woman still in high school ended up in charge of the 24-hour security. Others facilitated the difficult meeting circles in the evening or helped with other aspects of the camp life. "Within two weeks they had become integral, fully participating members of the Peace Camp, taking on major responsibilities."

Dana Kagis says that for her the camp was "an exercise in community." There were problems but there were also profound rewards. "Ecofeminism generated some backlash and some men who I would call strong feminists held workshops on unlearning sexism. I think they handled it extremely well."

On the last morning of the blockade, Mable Short, a matriarch of the Nuu-chah-nulth, announced a call to the chiefs to reclaim native land in the Clayoquot, and asked to be the first arrested that day.

Near her, standing in the road with others who were prepared to be arrested, were two twelve-year-olds. "I was willing to be arrested because it would mean actually saving a tree," said Eva Prevost. Her friend Elizabeth Reed, whose native name means Black Bear, agreed: "I wanted to save a tree." "I wouldn't get arrested for anything else," said Eva, "but it's better than being arrested for criminal reasons."

In the end the logging trucks stayed away. Mable and her sons retreated ceremonially to their car and other blockaders broke into celebration. In the line-up for supper the night before, in the midst of hugs and advice and listening, Jean McLaren had declared "I've just facilitated my last work-

shop at this camp. Tonight I'm going to have my face painted." When the sun rose at the Kennedy Lake Bridge, there she was, with a pink and blue striped face, dancing up a storm.

There could have been bloodshed this past summer. The *Times-Colonist* compared the blockade to the Winnipeg General Strike of 1919, the FLQ Crisis in 1973, Oka in 1990. The difference, points out Valerie Langer, is that this one was peaceful.

And the reason for that is women planned for it to be peaceful and worked very hard to keep it that way.

—***Focus on Women*, November 1993**

◆

WOMEN'S STATEMENT: We are from the group of women who were arrested together on "Women and Children's Blockade Day," July 21, 1993....

On July 21st we individually stood up in our belief that women have a special voice for the environment....

The court is the product of a world in which corporate dollars and powerful institutions collude in the destruction of our environmental integrity....

As women, we stand before you with complete awareness that you do not reflect nor entirely understand our values, ethics or basic approach to our world. We have watched as Sile (Sheila) Simpson was sentenced to six months in jail for opposing clearcut logging in the Clayoquot region, and compare this to the four-month sentence handed to William Seufert for sexually assaulting his two step-daughters from 1985 to 1991, starting at the time when the youngest daughter was a mere seven years of age.

In sitting in this courtroom, we grow weary at having to sit in subservience to male-defined language and rules of order. We are enraged that the court has the power to imprison us...for the supposed crime of opposing the ongoing systematic environmental desecration of our planet....

Throughout history we as women have been robbed of our power....

We are painfully aware of the court's seeming inability to set aside its double-talk and legal protocol. We plead with you in good faith: The planet is in peril, do not hastily judge us for our actions; *please, listen to our words.*

—Women's statement to the court

We organized training in nonviolence every day. We developed the circles as a way of decision-making, and worked on how they could be facilitated ...We based our actions on the Women's Pentagon Action and the Women of Greenham Common.

RESPONSE TO THE PROTEST

A 10-week-old lamb was shot sometime late Sunday or Monday and left to die a slow death in front of eco-forester Merv Wilkinson's home.

With three lambs deliberately killed since late February and six incidents of vandalism on the property, Wilkinson is beginning to suspect a vendetta....

[Clayoquot arrestee] Wilkinson is known for his opposition to clearcutting and preference for selective logging, which he applied on the 200-hectare property at Cedar he manages. The site is regarded as a model for eco-forestry.

The sheep are used to weed the underbrush.

—Judith Lavoie, ***Victoria Times-Colonist***

◆

Peggy Fraser, a 15-year-old half-native girl, was hit by a pick-up truck at the Peace Camp in Clayoquot Sound last October, on one of three weekends the camp was invaded or attacked by local youths.The man who drove the truck, Richard Brian Lee, was questioned by the police. No breathalysers were administered. No one was charged with hit-and-run. The RCMP recommended [a charge of] dangerous driving causing bodily harm. Steve Stirling, the Crown prosecutor, charged him with a relative slap on the wrist offence of driving without due care. The first charge is criminal. The decision was referred to senior Attorney General's Ministry officials. Stirling says the proper charge should have been hit-and-run.

Peggy's mother is from the Clayoquot and Ahousat bands. Peggy Fraser has the right to an impartial justice system that will prosecute anyone who assaults them.

—Sid Tafler

◆

Protester Bonnie Glambeck said the police were slow to respond. On Saturday about 2:30 a.m. six men, identified by camp people as Ucluelet residents, verbally harassed people at the camp's main gate, yelling threats and obscenities.

Glambeck said one man threatened to shoot them if they contacted the police. A woman alone in a tent felt terrorized when her tent was hit with rocks.

Some of the same people showed up the next night. A woman came to the front gate at 3:00 a.m. to create a diversion while other people broke through the woods into the camp. They yelled threats and threw rocks at the tents.

—Denise Helm, *Victoria Times-Colonist*

◆

Canada's ambassador to the U.N. for the Environment, Arthur Campeau,...claimed the [Clayoquot] blockade was making Canada look like an "environmental outlaw" and said, "It is causing us damage greater than we are prepared to admit.... It is time we put a stop to this damage."

—quoted by Joyce Nelson, *Canadian Forum*, July/August 1994

◆

A province-wide poll of 500 people found...a total of 81 percent...said the logging challenge should be expected because of poor company practices that have earned public distrust...62 percent said the United Nations should be asked to supervise logging.

—Robert Williamson, *Globe and Mail*

◆

We cannot have children disagree with authority now, can we...they might [not] grow up to be docile, obedient citizens...to respect the authority of greedy corporations...to pollute and obliterate the remaining resources of the planet....

—David Nicoll

SAVE WHAT'S LEFT

4

DISORDER IN THE COURT: MASS TRIALS—1

THE MASS TRIALS OF THE Clayoquot protectors began in confusion and continued that way through to the end. Defendants were often unsure of their rights, unable to secure legal representation, and confounded by the process itself which was unusual, to say the least. Adding significantly to this unfortunate state of affairs was the perception of injustice

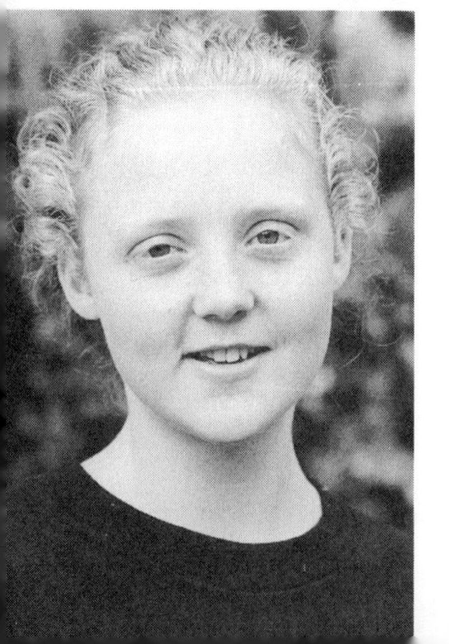

and corruption surrounding the case.

TIPPING THE SCALES OF JUSTICE

Like a giant locomotive pulling 50 cars from a standing start, the first mass trial of Clayoquot Sound anti-logging protesters charged with criminal contempt began slowly Wednesday in B.C. Supreme Court.

Crown counsel Brian Rendell got the prosecution's case underway shortly before 2:45 p.m. after Justice John Bouck spent the morning rejecting more applications for adjournments, one for severance from the 49 other accused, and protester Margaret Ormond's application for a jury trial.

The trial was to have started Monday, but Bouck allowed a two-day adjournment so people could seek counsel. Bouck rejected a preliminary objection raised by lawyer Robert Moore-Stewart, who said the B.C. Attorney General's Ministry should not be allowed to continue to prosecute because of a conflict of interest over MacMillan Bloedel, the forest company with timber rights in Clayoquot Sound.

Moore-Stewart pointed to the province's multimillion-dollar purchase of MacBlo shares in February and said B.C. was the largest single shareholder in the company. Once the proceedings began, 41 of the 50 on trial entered not guilty pleas. Warrants were issued for the arrests of Lance Davis, Leesa Heyward, Shawn Parkinson, Eric Priest, Jonathon Pulker and Todd Andrew Richer, all absent from court.... Tzeporah (Suzanne) Berman has her case severed and a trial set for December 5.

—**King Lee,** ***Victoria Times-Colonist***

◆

[Judge] Hutchison acknowledged the genuineness of the environmental views of the defendants and their unanimous indication that their action was not directed at the court but at MacMillan Bloedel. But their conduct in law was criminally contemptuous of the court injunction against interference with MacBlo's logging rights. He also said the defendants were "exemplary in every way," in the courtroom and were respectful and polite.

—**Roger Stonebanks,** ***Victoria Times-Colonist***

◆

The important thing to remember here is that the RCMP are complying with the wishes of an agent who represents a very large multinational....

Robert Hunter says "It's the end of British Columbia. Industry, government, labor, now the judiciary—all linked to meet their own short-term ends. B.C. has learned nothing."

—Lon Wood, ***Victoria Times-Colonist***

◆

One defendant expressed his bitterness at the help given MacBlo by the police:

Here are a series of questions I asked Lorne Dixon [of MacMillan Bloedel, during cross-examination at the mass trials]:

1.Upon picking up the envelope [from Sgt. Doyle at the Ucluelet police station], did you look inside of it to make sure it contained Polaroids having to do with the arrestees on, for instance, the 16th of August?

Answer: No.

2.Then you knew what the envelope would contain, namely the Polaroids?

Answer: Yes.

3.Did you make the request for the Polaroids from the RCMP detachment in Ucluelet?

Answer: Yes.

4.So the RCMP, in effect, was acting upon your request to turn over the Polaroids? Is that so?

Answer: Yes.

5.Did the RCMP comply quite willing?

Answer: Yes.

The important thing to remember here is that the RCMP are complying with the wishes of an agent who represents a very large multinational.... Later, I asked Dixon where he had gone upon leaving the Ucluelet Station with the Polaroids. He made the admission that he went to a restaurant. He did so rather proudly, laughing about it....

I asked him if he took [the photos] out of the envelope while he was in the restaurant, and he replied that he didn't....

Under further questioning by me, Dixon maintained he had kept the Polaroids and appended information in the strictest confidence. He did not remove them from the envelope while in the restaurant and had sub-

sequently taken them to MacMillan Bloedel headquarters in Ucluelet where, when he was not examining them, they were kept in a vault.

When Richard Bourne [process server for MacMillan Bloedel] got up on the stand and I got my chance to cross-examine him, I took him to the same restaurant Dixon had gone to and asked him if he had ever seen Dixon remove the Polaroids from the envelope while he was there, and Bourne testified that Dixon not only removed them from the envelope but passed them around the restaurant.

—Ernest Hekkanen, *Everwild* newspaper, January/February 1994. (Ernest Hekkanen is a Vancouver artist and writer who has had two books published.)

◆

The mass trial process is an obstruction of justice. With no specific dates set, people cannot call witnesses or have a lawyer on hand unless they sit there for a month.

—Andrew Gage, University of Victoria

JURY TRIAL

JURY TRIAL NEEDS CONSTANT DEFENCE: A right to humane and rational assessment of controversy was pretty dim until jury trial was offered by the court of Henry II....

The jury played an important role in the democratization of English society. It became the leveller of the rich and the poor, the powerful and the helpless. It was a little parliament whose constituents were the entire public....

The jury epitomized the proposition that law flows from the people.... Growth was not smooth. The role of the jury as the provider for justice for the common citizen was opposed by the powerful in the 17th century much as it is today....

The Crown fought the trend of jury independence by such acts as removing cases to the infamous Star Chamber or by careful screening of jurors.

Nevertheless, judicial decisions and legislative acts continued to entrench the principles of individual freedom with the right to trial by one's peers....

Then as now, the jurors were the forerunners in setting current community standards, where the courts without juries lag behind hobbled by isolation....

Disorder in Court: Trials — 1

It was at this time that jurors, and not legislators, were spearheading the rights to freedom of speech and assembly and freedom of religion, freedom from self-incrimination and freedom of the press that our ancestors brought to Canada and the United States; the principles which they claimed as their birth right.

—R.F.M.

◆

WHY A JURY: The blessing of a jury decision is that the jurors reflect a cross-section of the community's concept of what is a fair award in the circumstances.... The decision of each man is necessarily different....

No matter how sincere he is in wanting to be impartial and objective, his thinking is the product of his past. He has the advantage, of course, that eight different people tend to correct one another's errors and more truly smooth out the rough edges of bias.

Sir Patrick Devlin said in his book *Trial by Jury* (p. 159), on the subject of "Jury as Safeguard of Independence and Quality of Judges:"

> "The first [purpose]...is that the existence of trial by jury helps to ensure the independence and quality of the judges. Judges are appointed by the executive and I do not know of any better way of appointing them. But our history has shown that the executive has found it much easier to find judges who will do what it wants than it has to find amiable juries. Blackstone, whose time was not so far removed from that of the Stuarts, thought of the jury as a safeguard against the violence and partiality of judges appointed by the Crown.' "

—From an article by Ronald F. MacIsaac written 20 years ago for the *Canadian Bar Journal.*

◆

The Charter states that a defendant may have the right to a jury trial if the crime with which he is charged is punishable by a maximum prison sentence of five years or more.... Six months is the standard for jury trials in the United States....

Referring to the length of jury trials,...it is also true that lynching takes only a 100th the time of a non-jury trial.

—Kirk Makin, *Globe and Mail*

The essence of what I have to say is that the existing forest management system is not working, it is dysfunctional and must change....

MacMillan Bloedel's Record

1. November 25, 1969:
CONVICTED for destroying fish habitat through negligent logging practices (2 counts). Fisheries Act, Section 30 and 33(2). Fined $750.00 on each count.

2. August 2, 1973:
CONVICTED for destroying fish habitat through negligent logging practices. Fisheries Act, Section 33(2). Fined $2,500.00.

3. April 25, 1975:
CONVICTED for polluting fisheries waters. Fisheries Act, Section 33(2). Fined $2,500.00.

4. April 5, 1976:
CONVICTED for destroying fish habitat through negligent logging practices. Fisheries Act, Section 33(2). Fined $1,500.00.

5. June 12, 1978:
CONVICTED for destroying fish habitat through negligent logging practices. Fisheries Act, Section 33(2). Fine not reported.

6. March 27, 1979:
CONVICTED for destroying fish habitat through negligent logging practices. Fisheries Act, Section 33(2). Fined $1,000.00.

7. January 20, 1978:
CONVICTED for destroying fish habitat through negligent logging practices. Fisheries Act, Section 33(2). Fined $1,500.00.

8. October 7, 1981:
CONVICTED for polluting fisheries waters. Fisheries Act, Section 33(2). Fined $4,000.00.

9. January 29, 1982:
CONVICTED for polluting fisheries waters. Fisheries Act, Section 33(2). Fined $10,000.00.

10. January 29, 1982:
CONVICTED for polluting fisheries waters. Fisheries Act, Section 33(2). Fined $12,500.00.

11. September 25, 1982:
CONVICTED for polluting fisheries waters. Fisheries Act, Section 33(2). Fined $10,000.00.

12. January 14, 1987:
CONVICTED for polluting fisheries waters. Fisheries Act, Section 33(2). Fined $3,500.00.

13. August 7, 1987:
CONVICTED for polluting fisheries waters. Fisheries Act, Section 33(2). Fined $6,500.00.

14. October 20, 1987:
CONVICTED in Ontario for water pollution. Environmental Protection Act, Section 13(1)(d).

15. October 20, 1987:
CONVICTED in Ontario for failing to notify authorities of pollution event. Ontario Water Resources Act, Section 16(3) and 16(4).

16. July 17, 1988:
CONVICTED for violating the terms of its pollution permit. Pesticide Act.

17. October 24, 1988:
CONVICTED for destroying fish habitat through negligent logging practices. Fisheries Act, Section 33(2). Fined $1,000.00.

18. May 31, 1989:
CONVICTED for polluting fisheries waters. Fisheries Act, Section 33(2). Fined $15,000.00.

19. June 3, 1990:
CONVICTED for polluting fisheries waters. Fisheries Act, Section 33(2).

20. June 19, 1990:
CONVICTED for polluting coastal waters. Waste Management Act, Section 44.

21. September 19, 1990:
CONVICTED for failing to report a chemical spill. Waste Management Act, Section 20.

22. October 23, 1990:
CONVICTED in Ontario for water pollution. Ontario Water Resources Act, Section 16(1).

23. November 15, 1990:
CONVICTED for violating the terms of its pollution permit. Waste Management Act, Section 24.

This list is not complete. Between 1991 and the present, MacMillan Bloedel has been in non-compliance with the law on at least 50 occasions.

MacMillan Bloedel, in our submission, was engaged in a continuous, reckless and flagrant disregard of Guidelines...

DEFENCE OF NECESSITY

Lawyer Paul Hundal was a tower of strength in the mass trials. He defended protestors at most of the trials, worked long hours, and freely gave of his time to the unrepresented.

Defence Counsel Paul Hundal, for Peter Rowan and Eric Priest, Defence of Necessity and Section 27 of Criminal Code:

Everyone is justified in using as much force as is reasonably necessary to prevent the commission of an offence...that would be likely to cause immediate and serious injury to the person or property of anyone. Rowan and Priest were aware of MacMillan Bloedel's extensive criminal record for violations of the Fisheries Act....

The [company is] at this very moment under investigation for further potentially indictable offenses.... Therefore MacMillan Bloedel, in our submission, was engaged in a continuous, reckless and flagrant disregard of Guidelines...and thereby was likely causing serious property damage....

◆

The protestors had engaged experts to testify as to the immediacy of the threat to life, and the necessity of protest. Here are portions of the Clayoquot Evidence Offered but Declined by the Court. The witness is O.R. Travers, Registered Professional Forester:

My immediate intent was to assist those persons on trial who were using the "defence of necessity" to defend their conduct in Clayoquot Sound this past summer.

The essence of what I have to say is that the existing forest management system is not working, it is dysfunctional and must change.... The present forest policy is concerned almost exclusively with the short term....

» There is no wildlife habitat legislation;

» Fisheries legislation is not designed to prevent damage, only used to prosecute offenders after the fact;

» Legislation is expected to exempt timber operations from environmental assessment;

» [Of the] Forests Act: ...judicial review is...required;

» The right to manage public foresters has been assigned to private corporations.... Staff to enforce the guidelines...is inadequate....

Sustainability...does not appear to be understood; ...management actions routinely put at risk non-timber values such as water quality and the continued existence of wildlife and plants that are closely associated with old-growth forests.

The intent of the forest management concept inherent in the forest policy to cut every commercial tree...[is that] no trees [younger] than the rotation age (say 60 to 120 years) [are cut]. [A] U.S. Forest Service study documented that over 600 species of plants and animals are associated with old-growth Douglas fir....

The provincial government is slowly responding to these issues.... The industry response has been to readily disperse operations into unlogged valleys...a high-grading of the forest, fragmenting...intact watersheds.

Many of the land use problems in B.C. forests could be resolved if provincial planning kept up to date with the emerging concepts in the forestry profession, especially the concept of ecosystem.... Ecosystem management is now being implemented in Washington State and the U.S. Forest Service in U.S. federal forests. Similar concepts and planning tools are yet to be put in place in B.C.

—Oliver Raymond Travers, RPF, B.Sc.F. (1966) University of British Columbia (Silvics Option); MF (1970) Oregon State University (Major, Forest Management; Minor, Silviculture)

INTERNATIONAL OBLIGATIONS

CARACAS DECLARATION: The Fourth World Congress of the International Union for the Conservation of Nature (IUCN) met in Caracas, Venezuela in February 1992. Both the B.C. Ministry of Environment, Lands and Parks and the Ministry of Forests have endorsed the principles enunciated in the Caracas Declaration.

It states: "Our natural wealth is being eroded at an unprecedented rate, because of...[among other things] excessive consumption of natural resources...so that the future of humanity is now threatened."

"The world community must adopt...sustainable use of the environ-

ment, and the safeguarding of global life-supporting systems." This document became law on December 29, 1993.

—Friends of Clayoquot Sound newsletter, Winter 1993/94

CIVIL LIBERTIES

The B.C. Civil Liberties Association is of the impression that the trials had always been attended by top lawyers. They must not have been present at the start of the mass trials when the lawyers chosen by the accused were unable to act. Here is an edited extract from deliberations on preservation of the old-growth forest.

The underlying cause of the preservation of an old-growth temperate rainforest has drawn worldwide attention. Yet the legitimate debate on this important issue has become entangled with a messy debate on the role of the courts in this essentially political process.

The courts have been attacked by the protestors, by provincial cabinet ministers and by the media. For the most part, these attacks have centred on the conduct of the trial, especially the courts' refusal to allow the protestors to argue that they have rights protected under the Charter of Rights and Freedoms to a sustainable environment, or that they acted out of the necessity to protect the environment....

The liberties issues which the court must be careful to protect are...the right to counsel, the right to a fair trial *on the issues*, and the right to a fair and unbiased determination of whether or not the protestors were validly arrested. Some of us would like to see the courts "overrule" the government on the issue of logging old-growth forests. But that is not their role in adjudicating a contempt charge; their role is to be fair....

Where do we go from here? The BCCLA has kept in close contact with the lawyers for the protestors, and has focused on the issues of the right to counsel and the mass trials. These certainly trouble us, but so far the trials have always been attended by very capable criminal defence lawyers, including some of the best defence lawyers in the province. We have watched carefully as the issues have unfolded, and we have discussed these issues at every Board and Executive meeting since June of this year.

We may decide to intervene at the appeal level if the civil liberties issues crystallize in a way where our input is useful; but we will not take sides on

"The world community must adopt...sustainable use of the environment, and the safeguarding of global life-supporting systems."

the underlying political issues, or ask the courts to decide those issues, any more than we did with the abortion protestors in 1989.

Because we, as a civil liberties organization, must keep a clear focus on the proper role of the courts, on due process for the accused, and on the limited role of the courts in political issues in a democracy, where the people—through their elected government—are sovereign.

—Andrew Wilkinson, President, B.C. Civil Liberties Association, in *The Democratic Commitment*, December 1993.

THE TRIAL

The hundreds of arrested protesters finally came to trial. The mass ceremonies were to be held in a theatre, but the first batch of 50 accused were called for a hearing in the courthouse before Mr. Justice Bouck, and the following document overleaf was said to be the complaint. It didn't make much sense to the accused, as it did not bear their names, and the plaintiff against them did not appear to be the court. The judge drew up what appeared to me to be an understandable document and gave it to the prosecutor, but it was never adopted. The accused were mainly unrepresented by any lawyers, let alone top lawyers, so confusion continued.

The weird and wonderful world of politics opened a window on the confusion as to who was responsible for charging the forest defenders with criminal rather than civil contempt. No one admitted responsibility.

Protestors and the public complained that their peaceful and civil protests had unfairly been upgraded to criminal contempt. This is how the press described it.

Senior officials of the offices of the Attorney General gave media interviews respecting the issue of who initiated and controlled the charges. This issue progressed to the point that on October 22, 1993, the Chief Justice of the Supreme Court of British Columbia took to his bench to make a public statement on the issue of who commenced the proceedings.

The Chief Justice stated: "The proceedings for contempt launched in July were begun and have been conducted by counsel from the Ministry of the Attorney General."

In his public statement the Chief Justice responded to public statements made by five elected members of the Legislative Assembly for the province of British

No. C916306

Vancouver Registry

IN THE SUPREME Court OF BRITISH COLUMBIA

BETWEEN:

MACMILLAN BLOEDEL LIMITED

PLAINTIFF

AND

SHEILA SIMPSON, ET AL.

AND PERSONS UNKNOWN

DEFENDANTS

NOTICE OF MOTION

IN THE MATTER OF THE PROSECUTION FOR CONTEMPT OF COURT OF JONATHON DAVID PULKER, WARREN GEORGE BELL, CHRISTOPHER DALE PHILLIPS, LANCE DAVIS AND MARGARET SCHMIDT.

TAKE NOTICE that the Attorney General of British Columbia will be assuming the prosecution of this matter as a CRIMINAL CONTEMPT OF COURT.

FURTHER TAKE NOTICE that in support of this motion will be filed the Affidavit of S/Sgt. Len Doyle.

DATED at the City of Victoria, in the Province of British Columbia, the 20th day of July 1993.

Brian Rendell—Crown Counsel

THIS NOTICE OF MOTION is issued by Brian Rendell, Crown Counsel, whose place of business and address for service is Crown Counsel Office, Third Floor, 910 Government Street, Victoria, B.C.

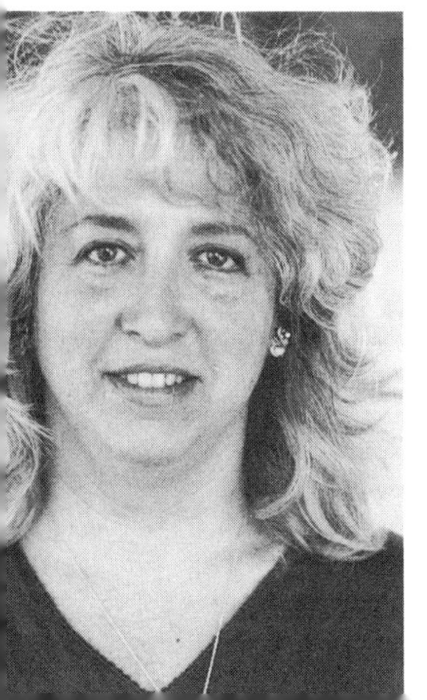

Columbia a few days earlier.... This group of five MLAs included four senior cabinet ministers. An extract from their letter follows:

AN OPEN LETTER TO CONSTITUENTS OF LOWER ISLAND:

October 18, 1993

Dear Constituent:

...The extent of public anxiety in our communities has caused us great concern as MLAs. We are particularly troubled by suggestions that the conduct of the trial and resulting jail sentences were directed by the premier or the Attorney General....

The decision to commence contempt proceedings against the Clayoquot protestors was a decision of the courts, not the government. Because of the independence of the judiciary, it is the courts that are responsible for commencing and conducting contempt proceedings, with the direction of the Attorney General or the premier....

We understand that many of our constituents want the government to step in to influence or discontinue these proceedings. However our laws require that the courts be free of political interference....

Sincerely,

Robin Blencoe, MLA, Victoria Hillside
Gretchen Brewin, MLA, Victoria Beacon Hill
Elizabeth Cull, MLA, Oak Bay Gordon Head
Andrew Petter, MLA, Saanich South
Moe Sihota, MLA, Esquimalt Metchosin

◆

Contempt of court is an...enigma. Clayoquot protestors are...lost in the puzzle, according to University of Victoria's law professor Hamar Foester.... Criminal contempt is...not explained and regulated in the law books and has no maximum penalty.

—James McKinnon

◆

OPENING DAY OF THE TRIAL BEFORE THE HONORABLE MR. JUSTICE JOHN BOUCK

Here are fragments from the daily records of the trial and its slow and tumultuous progress.

THE SHERIFF: Order in court.

THE CLERK: In the Supreme Court of British Columbia, this 1st day of September, 1993, in the matter of MacMillan Bloedel versus Sheila Simpson et al. at bar. My Lord.

PROSECUTOR, ROBERT GILLEN: My Lord, Gillen, initial R, appearing for the Crown. With me is Mr. Brian Rendell.

DEFENCE COUNSEL, RONALD MACISAAC: May it please you, My Lord, I appear for Stuart Parker et al.

DEFENCE COUNSEL, JAMES MILLER: My Lord, I appear on behalf of Betty Krawczyk and Yvonne Kato.

DEFENCE COUNSEL, JAMES HELLER: May it please My Lord, I act on behalf of Andrew Stevenson, Warren George Bell, Margaret Schmidt and Jennifer Maxwell....

◆

UNIDENTIFIED SPEAKER: My Lord, I submit that your words in response to lawyer Mr. Orris' plea for an adjournment on Monday...many of us here are finding ourselves in a situation without legal counsel because of the booking. Our lawyers can't make themselves available.... I spent one day in the public library, which is not quite enough to prepare myself for the complexities of law to defend, particularly on the question of *mens rea*, which is my intent at the blockade, so I also appeal for an adjournment, at least one week.

THE COURT: I cannot allow the adjournment on that basis. We have all been through that argument about no lawyers. I gave you some time. There's been—not an enormous amount of time, but substantial amount of time available to get counsel. There is no assurance you will get counsel if—even if—there's an adjournment, so we are going to proceed. Well, let me just say something about that: you know, the Criminal Code doesn't cover these proceedings, so we are sort of allowed to make the rules as they

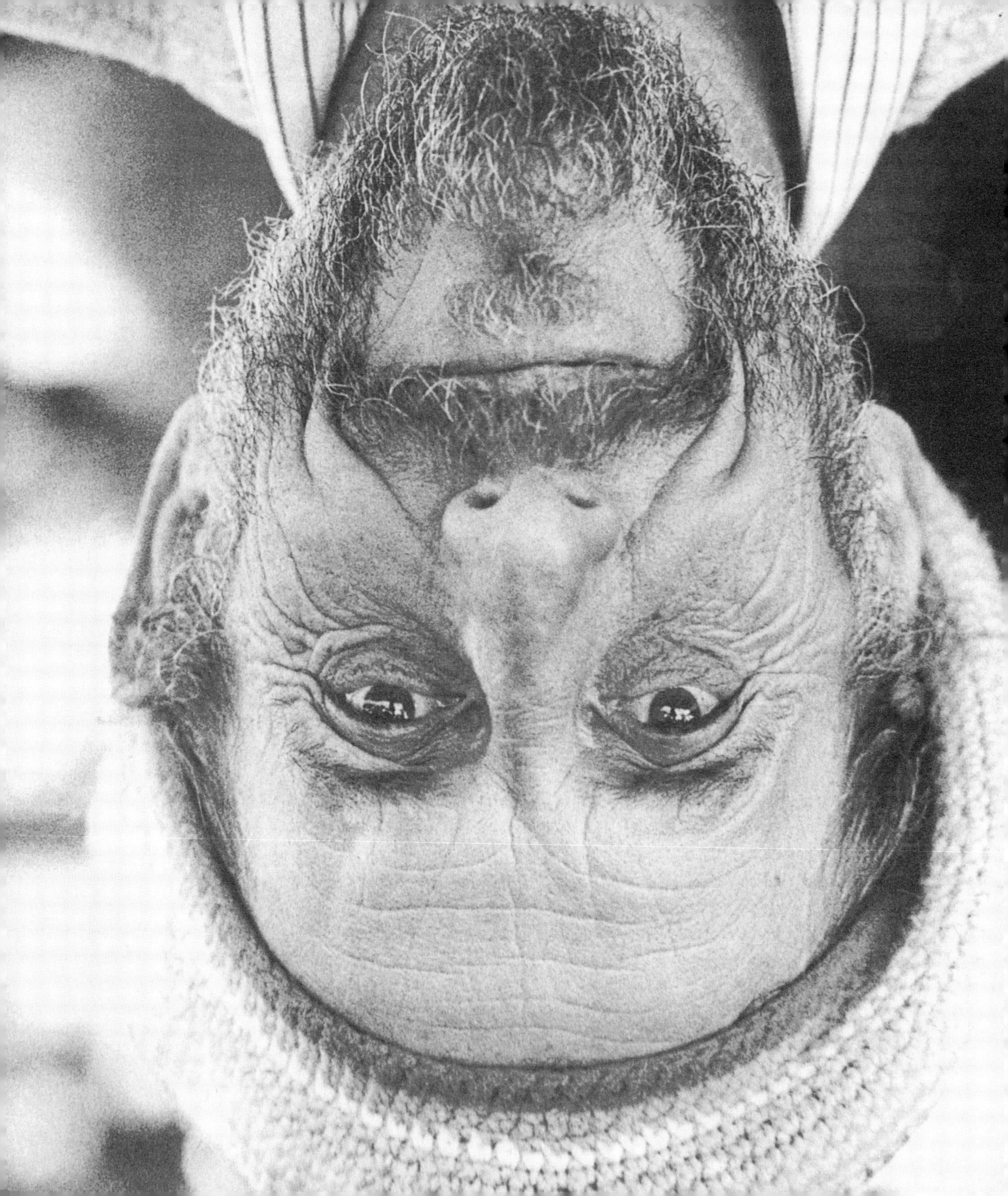

...many of us here are finding ourselves in a situation without legal counsel because of the booking. Our lawyers can't make themselves available....

go along where they are reasonable and fair as much as we can....

◆

MARCELLE BODMAN: I would like to put forward a preliminary motion that my case be severed as I am an Australian citizen and I have no immigration counsel. I have not been able to find anybody as yet, I also have not been permitted to work. I have been detained by the court to appear today and I have been waiting around for the last month-and-a-half and I have no money and I have been living off the graces of friends....

BETTY KRAWCZYK: My Lord, Judith and I have been incarcerated at the Burnaby prison and also held at the Victoria city drunk tank, and if we have to come back on Tuesday, could we please be returned to Burnaby? We have already been here five days and it is horrendous.

THE COURT: Ma'am, I am sorry I do not run the prison system. I cannot tell them where to put you and they have beds available in one place and not another, so it is up to them to run the prison system, not up to me.

ANDREW SWAIN: My Lord, my name is Andrew Swain. I am 18 years old and still attending high school. I have absolutely no knowledge of Canadian law, criminal or civil, and I believe that it would be a travesty that would reflect very badly on this court and indeed on the citizens of B.C. for me to be tried in this manner. I have made every attempt to contact counsel. I have contacted four Victoria area lawyers and have failed to get through to a fifth.

These trained people are all interested in the case but unfortunately cannot be ready in the time allotted between August 30th and September 1st, to be able to properly prepare for a defence or clear their timetable—now, I would ask, with Your Lord's pleasure, to please have my trial adjourned for five weeks so that I may be able to study the law myself or obtain counsel who would do this for me.

THE COURT: Thank you, sir. I am sorry, I ruled on the adjournments. There is not going to be any adjournment. We are going ahead with the trial.

UNIDENTIFIED SPEAKER: My Lord, I would like to make a preliminary motion as to the expeditious nature of these trials. There is a considerable amount of pressure on the court to speed us through and this fast-tracking, I think, shows a certain bias to pressure us to plead quickly, to defend our-

selves quickly and get through. Murderers and rapists and other groups have years to prepare for their trials and I do not think we are receiving the same rights. As well, MacMillan Bloedel has 50 outstanding charges against it. They have basically dirty hands and they are proceeding against us. Basically I would like the same opportunities to prepare as these other people do....

ANDREW SWAIN: I am attending high school at Oak Bay. On September 7th I intend to claim my right to an education as a Canadian citizen and on that day and every day thereafter I shall be attending school at Oak Bay High and if the Crown intends to send around people to arrest me they can do so, which will involve the removal of me from my classes.

THE COURT: I suggest you not do that. I know it's not convenient to you, but I have no alternative but to issue a bench warrant and hold you in custody during the whole length of the trial....

Editors' Note: Andrew Swain was jailed for several days before agreeing to attend court.

UNIDENTIFIED SPEAKER: My Lord, before the opening submission I have a question I would like to ask about the witnesses that we'll be allowed to bring in as a defence point of view.

THE COURT: Well, I see I am in a difficult spot. I want to make sure that the people that are here who don't have lawyers, I give them as much assistance as I can. On the other hand, I am supposed to be impartial between the Crown and yourselves and I am not supposed to be giving the Crown any assistance, so all I can say is that you need to read the rule and see what you have to do to issue the subpoena and take such proceedings as may be required.

UNIDENTIFIED SPEAKER: Thank you, My Lord. My Lord, where can we read these rules?

THE COURT: They are in the law library.

ANDREW SWAIN: My Lord, with what time do we read these rules?

Editorial note: The arrestees were required to be present in court.

THE COURT: I don't know.

◆

I believe that it would be a travesty that would reflect very badly on this court and indeed on the citizens of B.C. for me to be tried in this manner.

SANDOR CSEPREGI: I stood on that logging road because it is my civil duty to inform the government that we are in a state of environmental emergency. With ten percent of the rainforest left, we must protect Clayoquot Sound; the clearcutting must stop. We, the people, will not stand aside any longer as multinational corporations like MacMillan Bloedel, with the consent of our government, continue to exploit our land for the purpose of economic gain. This is an issue between people and logging practices of MacMillan Bloedel, and it is an issue between the people and the NDP government. Your Honor, you have a choice. I would like to ask you to dismiss this court, to stand down, to stand with us, join us for Clayoquot Sound. I cannot stand in court. It is unjust. I will not stand in court.

THE COURT: There will be—call up some help and we will place him in custody downstairs.

UNIDENTIFIED SPEAKER: Your Honor, I would suggest that many of these proceedings have already been interrupted so many times—the Crown's proceedings—that I cannot believe we are still here and I would recommend that if this individual is being incarcerated right now for standing up for his beliefs that at least he be given the comfort and the right to a defence.

THE COURT: He has a right to a defence.

UNIDENTIFIED SPEAKER: Does he have a right to a lawyer?

THE COURT: If he can afford one.

UNIDENTIFIED SPEAKER: My Lord, on the question of right for legal representation, this court denied flatly the right for 30 days adjournment so that our lawyers could clear their schedules.... I have just heard that the case on the other 50 was just adjourned in the interest of the prosecution. I believe that really shows, My Lord, in all due respect, that this is not a fair trial, this is clearly weighted in the interest of the prosecutor....

ROBERT MARK OSLEEB: My Lord, I would like to make an application for trial by jury....

DAVIDA SEFERITH: My Lord, I would like to make an application for trial by jury....

THE COURT: Well, I think I have heard all those and I think I have told everybody that there is not going to be a trial by jury in these proceedings

because the law says that is not the way to try these matters. Now I'll accept—take it for granted that everybody has a note—a continuing motion that they want a trial by jury so you don't have to keep making one.

◆

SEPTEMBER 8, 1993: SECOND WEEK OF THE TRIAL

RON ASPINALL: Yes, My lord. What is the method of subpoenaing witnesses: Could I ask the court to do that...?

THE COURT: I cannot do it, so you will have to read the Rules of Court to see how they provide for the subpoenaing...of witnesses.

RON ASPINALL: I have been told that a judge can order that.

THE COURT: I can in certain extreme emergency situations, but I am not going to in this case.

RON ASPINALL: Where do I go to get that information?

THE COURT: You'll have to read the Rules of Court.

◆

RON ASPINALL: Respectfully, the reason why I am asking about the Grannies is that they have not been in prison for a couple of days; they have been in prison over two months, which is longer than the expected penalty for civil contempt, and this is a major hardship. I am actually going to spend this lunchtime making a complaint to Amnesty. It is a problem that I think Your Lordship could help [with] in making a non-specific order that these women be transported to a normal jail for the upcoming weekend.

DAVIDA SEFERITH: My Lord, I believe there are other options for holding these women, and containing them in the fashion that they are being contained in the drunk tank is an infringement of their rights and freedoms under the Charter.

BETTY KRAWCZYK: My Lord, we have a complaint. We have had no personal supplies for 12 days. We are locked in a drunk tank. People come in there, they are drunk. They are raving. We have no personal toiletries. We cannot use makeup, we cannot keep a comb, we cannot do any of the things that make us feel like decent human beings when we come to appear in court.

The men's drunk tank is right down below us and they scream and hol-

ler and rattle the bars all night. Nobody sleeps. The pepper spray is often used. It was used in our cells two nights ago and when the pepper is sprayed, everybody is sprayed.

When people smoke in these cells everybody smokes, and just trying to get a breath of fresh air is a big production, and these are inhuman conditions to keep anybody for any period of time, Twelve days is just totally unreasonable. Thank you, My Lord.

THE COURT: Well, you have the option to sign a document in which you agree to obey the law.

BETTY KRAWCZYK: That is no option, My Lord.

THE COURT: Well, it may not be an option to you, but it is the option the law allows you and if you choose not to elect to take advantage of that option that is the end of it.

◆

ROBERT MAHER: My Lord, my application would be for severance so I do not prejudice other people's trials.

THE COURT: Gosh. How many times have I told you that I know there is a perpetual motion for severance. You just can't stand up and say, I want severance. You have got to stand up and show me case law and everything that supports an order for severance.

◆

SEPTEMBER 13: WEEK 3

RCMP SGT. DOYLE: Witness for the plaintiff, cross-examined by Mr. Moore-Stewart.

ROBERT MOORE-STEWART: Why did you provide people's names and personal addresses and dates of birth to MacMillan Bloedel?

RCMP SGT. DOYLE: Well, I cannot answer why. We did it as a—they provide us video tapes and photographs, we provide them with the polaroid.

ROBERT MOORE-STEWART: Why?

RCMP SGT. DOYLE: Why did they want it?

ROBERT MOORE-STEWART: No, why did you provide it to them? Why did you supply this personal information, dates of birth and addresses, from the arrest information?

I stood on that logging road because it is my civil duty to inform the government that we are in a state of environmental emergency. With ten percent of the rainforest left, we must protect Clayoquot Sound; the clearcutting must stop.

RCMP SGT. DOYLE: Because it was information they needed and they had information we needed; it was an exchange.

ROBERT MOORE-STEWART: How often is it that you open up your police file to a private party like this in a civil litigation...?

RCMP SGT. DOYLE: I would say it is probably the first time I have been involved in this situation but it is probably something that happens.

◆

SEPTEMBER 14

BETTY KRAWCZYK: I think I am the only one who made a public protest from Ucluelet, I was treated as a—a pariah in the Ucluelet police station....

In reading from this book on the Constitution, if you will bear with me, it says: "Everyone has a right on arrest or detention to be informed promptly of the reasons therefor," and on this form, Sheila Simpson et al., it says: "Take notice that the Attorney General of British Columbia will be assuming the prosecution of this matter as a criminal contempt of court." Is that what this means?

THE COURT: I am not here to answer questions ma'am. There it is.

◆

WITNESS: SGT. ROGERS, cross-examined by:

JANINE TOMANEY: I was held all day in Ucluelet while waiting to be transported to Nanaimo. Was I courteous and respectful? Was I misbehaving? Did I use foul language or anything like that?

SGT. ROGERS: Like I mentioned before, My Lord, any dealings I've had with the people from Clayoquot bridge area, I have never had any problems with any one of them.

BETTY KRAWCZYK: And you are aware that there was a real potential for violence there on the part of the loggers, that there were a lot of jokes and rumors circulating in town that the loggers were going to turn the Black Hole into the Red Hole, meaning that they were going to make the Black Hole run with blood? Did you hear those?

RCMP SGT. B.D. JOHNSTON: No.

BETTY KRAWCZYK: Do you remember reading us our rights at all?

SGT. JOHNSTON: No, I do not.

BETTY KRAWCZYK: I do not remember anybody reading us our rights, sir, and I don't believe they were read to us, and there was no one else that was arrested with us that I have talked to to date remembers it. Is it possible that we were not read our rights at all?

SGT. JOHNSTON: I do not know. It's possible.

◆

The trial was tumultuous, feelings were frequently vented. There were tears, cries, murmuring, laughter, even applause. A mass of people, without lawyers, in court for the first time, was a recipe for mistrial. Every spectator seat was in demand. There were many memorable speeches. I will start with that of my client.

ROBERT LIGHT: I am a master of fine arts graduate and I have lived and worked in Vancouver for the past nine years. I am single and have no dependants. I have lived my whole life in Canada. I come from a happy home. My parents brought us up in a good and supportive way. My mother is a retired nurse and my father is now retired too. He was a sergeant major in the RCMP. My uncle was a thirty-year veteran in the RCMP; he is now a lawyer and his ambition is to be a judge. My brother is an undercover officer in the RCMP.

My sister is a police officer of ten years in the Calgary police force. I have two cousins in the RCMP. Law and order is in the blood of our family. We serve and protect the values we treasure. I have worked to serve and protect the planet. I would like to examine some extenuating circumstances about my actions on July 9th, 1993.

It was my first day at the bridge. I did not contact the media nor did I speak with them. My action was directed totally towards MacMillan Bloedel in the spirit of nonviolence. I accepted the paper injunction from Mr. Bourne, I read it and put it into my pocket. I was there to nonviolently risk arrest but not to break the law for even if I was arrested you have not necessarily broken the law. Only a court can decide that.

I would challenge the moral legitimacy and constitutional legality of the injunction by this trial. I complied with the arrest officer in accordance with our peaceful direct action code:

> "Our attitude is one of openness, friendliness and respect towards all beings we encounter. We will not use violence, either verbal or

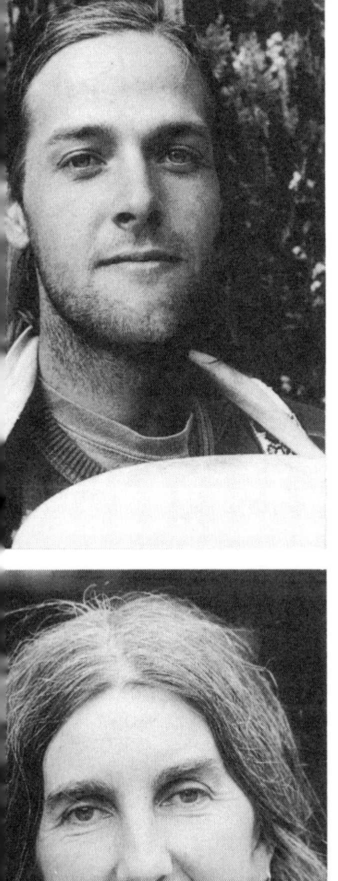

physical, towards any being. We will not damage any property and we will discourage others from doing so. We will strive for an atmosphere of dignity. We will carry no weapons. We will not bring drugs or alcohol. If you cannot accept our code of nonviolent action we cannot accept your participation in any of our actions."

I signed the undertaking and have honored my promise to avoid further arrest. I was arrested, I might add, when the charge was only civil; now it is criminal. Such retroactive action seems hardly just to me. As I said, my purpose for being here is to serve and to protect my planet. No one can take the law into his own hands but no one is above the law. I agree. We, all of us, are the law. To set yourself above the law is to set yourself above others. If we are ruled to have broken any laws we must obey the rule of law and welcome its punishments and its rewards.

I upheld the letter of the law and the spirit of justice by standing on the line. We had to think of the repeated violations by the forestry companies against the federal Fisheries Act and think about the loss of livelihood, the history of the conflict of interest and other laws violated by them, the effects of downstream pollutants due to clearcutting on the law of the sea.

There is international law and there are UN provisions of environmental protection. These just laws and others are served every time we stand on the line.

I believe in the law. Any law that serves justice. It was my intent on July 9th to embrace the law and challenge its truth in the court. When I was arrested I complied with the officers' wishes; unlike anti-environmentalists who used violence against us, we do not run from the law, we run with it.

In Plato's dialogue called Crito we have an ancient case authority on fidelity to the laws, even when and especially when they may be unjust. Socrates was imprisoned, condemned to die. Crito came to him with a plan for escape but Socrates would have none of it:

> "Imagine," Socrates said, "that I was about to make my escape when suddenly the laws of the city stood before me and demanded, 'Tell us Socrates, exactly what you are intending to do. Are you not planning to destroy us, so far as lies in your power?' I might answer, 'The government has injured me by imposing a sentence that is wrong.'"

"You could indeed say that," Crito interposed.

"The laws might continue, 'Your parents were married according to us, the laws of Athens, and under us you were well educated. In a sense, you are our child, as your father was before you. Now you are intending to run away like a slave. If you escape now, you will prove yourself a corruptor of law, and therefore likely to corrupt not only young men, but those of every age. You will violate the most sacred laws just to live a little longer. If you stay, you will die as a sufferer, not a doer of wrong, a victim not of the laws, but of men. Listen to us, Socrates, not to Crito.' These are the words I keep hearing, Crito," said Socrates, "and I can think of nothing else. Anything you may say contrary to them will be said in vain."

Sadly Crito said, "No, Socrates. I have nothing to say." "Then Crito, let us accept the situation, since this is the way God leads."

Not all laws are just. Hitler and Caesar gained power by wholly legal means. Gandhi said, "It is as important to disobey unjust laws as it is to obey just laws."

Not all laws are just. Hitler and Caesar gained power by wholly legal means. Gandhi said, "It is as important to disobey unjust laws as it is to obey just laws."

Think of the underground railway—Harriet Tubman, Sojourner Truth and the others. They broke the fugitive slave law. Now they have statues and statutes named after them. We have come here to have our actions judged legal or illegal. We too have a court, the court of public opinion. We want the public to judge for themselves our love of law.

Whether it is risking our lives for harpoon-hunted whales fifteen years ago or risking our freedom now for the forests, the law of love remains the same. This is what we obey, for without love we can never succeed; with love we can never fail.

...I truly believe that Judge Bouck is seeking his best understanding of justice of the written law. We may disagree on our interpretations of written law. I know that we, the protectors, seek justice in unwritten laws and laws of justice yet written for peoples yet unborn, with accents yet unheard. The great guardians of life in the rainforest have no voice but ours in this court.

These giant trees have stood and sustained other lifeforms, including us, for ages uncountable—the largest living beings evolution has ever witnessed on this planet. Now we all stand on guard for them. The price we pay is small, so small, for the debt of gratitude we owe them. I would be

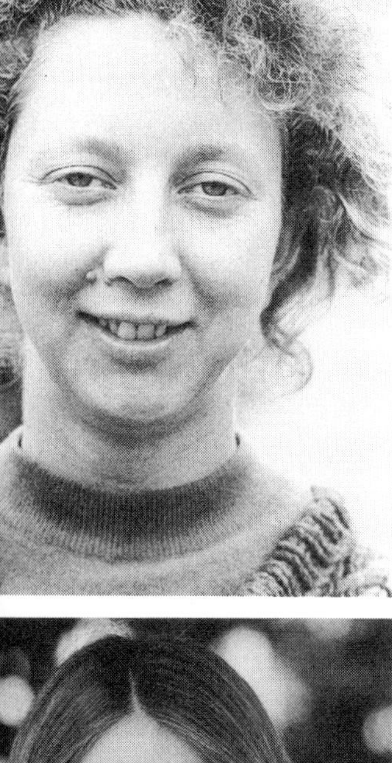

guilty of negligence, criminal indeed, if I did not do everything nonviolent I could to save them as they have so often saved me. I do not expect chaos to be borne from our protests but I do expect the birth of a new order.

We must remember how five hundred thousand men, women and children followed Gandhi into the Indian Sea in 1930 to break an unjust law and make salt. A needless and bloody revolution was prevented and a nation of free people was born. Mob rule is best prevented by nonviolent action. We have mob rule right now in the NDP government. They know representative democracy is neither representative nor democratic. In the last election twenty-three percent of the total public voted, less than fifty percent voted for the NDP; seventy-five people out of 3.5 million people in British Columbia now stand in a room to make legislation and laws for everyone.

"Representative democracy" is a contradiction like "military intelligence" or "sustainable development." Most people no longer vote. Five percent of Canada owns over 40 percent of its property and that's a fact from Statistics Canada. And this is the way of the whole world. We are living in this first global economic dictatorship and it's all built on a lie.

My intention at the bridge was self-evident. I questioned ownership. No one owns the air, no one owns the sea and I would put it to you all that no one owns the Earth. It is we who belong to it and I acted on this intuitive truth and stood on the road to destruction. The web of life versus the web of lies called ownership. And I have to ask who caught whom.

I would say that you have us right where we want you to have us. We have arrested an unjust practice and put it before the court of public opinion and there sat the media, ironically in the jury box, to witness it and take it to the people. And there is no appeal above, beyond, or without the people. We have brought a violent crime called clearcutting into worthy disrepute.

Those in MacMillan Bloedel who wish to abuse the law for their selfish ends and to obstruct the justice that we serve have fallen into a web of their own weaving. They are on trial on every medium and keep losing more every time we suffer more.

Our message is very simple: tell the men who sent you we have nothing to lose and a whole world to save, and tell the men who sent you that we

will never give up. By our strength of feminism, consensus democracy and nonviolence we can do more than save the trees. We may be planting the seeds of a new civilization. This greater trial is a triumph of democracy.

It is a strength of the many against the few, for the rich are the only minority. It is a wilful trial between public good and private greed, between ecosystems and ego systems, between the empowerment of justice and the enforcement of injustice, between life and death, between good and evil, between right and might. And it is between a post-violent vision struggling to be born in this world and a very violent world trying to kill it. It is a trial of truth between the chalice and the blade.

In Clayoquot Sound we are meeting legalized immorality with an outlawed morality, and we are winning. As much as we suffer, we are winning. And we are condemned to succeed. Not today or tomorrow but sometime soon Clayoquot Sound will stand alive and stand free forever. Years ago this movement began by liberating humans, then animals, and that continues on. Now we are in a struggle for total ecological liberation.

The seed is planted. Justice will blossom and its fruit will be in peace. A peace for all times, all places and all species. The struggle will see violence, not by us but upon us. History tells us patriarchy has only two allies to keep itself in power: hierarchy and militarism. Patriarchy is a relatively recent dysfunction, maybe six or seven thousand years old. In our struggle, the few good laws we do have come at a hard cost and the vast majority of laws in this culture still deal with the protection and the extension of private property. The indigenous people who lived here in symbiosis with nature for thousands of years had laws mostly dealing in sacred traditions and we still have a lot to learn from them.

Gandhi said that the powers that be will use any means to prevent justice. First they will ignore us, then ridicule us, warn us, threaten us, repress us, and then accept us. The struggle with love for our faith tells us our courage to suffer repression is stronger than our tormentor's will to inflict it. Consider the early Christians: the victims of the coliseum inspired others to conquer an empire with love and though their power of love was corrupted into the love of power, it was a great thing.

Finland in 1905 won her freedom from Czarist Russia with a six-week

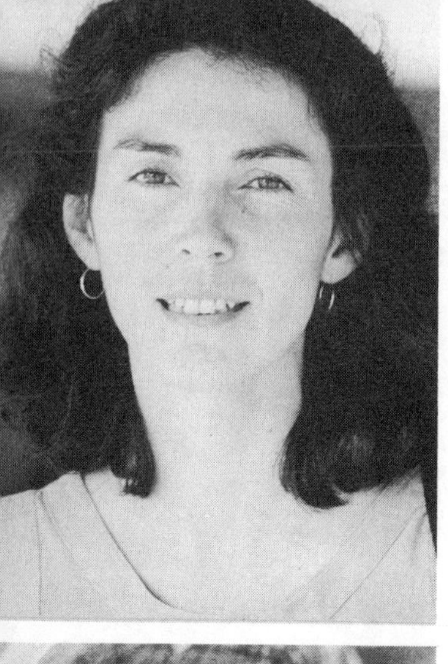

general strike. The garrison began to starve, they got on the railway and they left; not a shot was fired. We must look at Gandhi and Martin Luther King. We owe it to them to build on their successes. We must remember the seal hunt. I am born in Newfoundland, my great grandfather was a whaler and a sealer and now the seal hunt is ended. Did the repression stop us there? I doubt that it will stop us on the West Coast.

The corporations who wanted this injunction will learn by our nonviolent example how they own nothing and owe everything to the lifeforms around them. Our struggle is to win them over. As Martin Luther King said, that will be our greatest victory. He also said we use love and nonviolence, not that we lack the courage for battle nor the genius for strategic command nor the resources for success. No, we use love for, despite the repression and hatred we face, despite all the evidence to the contrary, we still believe and will always believe in our humanity.

Editor's note: Mr. Light, like others following him, chose jail rather than accept bail pending appeal.

◆

JANE SAVILE: ...The government has allowed multinational corporations such as MacMillan Bloedel the rights to rape and pillage our forests. What power do the people really have? What legal, democratic options are left?

Civil disobedience is the refusal to obey certain government laws or demands for the purpose of influencing legislation by nonviolent public actions.

At no time did I or any of the Clayoquot protectors show any disrespect towards any person or property. Our silent protest was directed at the government for allowing the blatant destruction of our last remaining temperate rainforest on Vancouver Island. What rights protect our environment, so crucial to human survival...?

It is time that apathy in Canadians ended. By co-operating with unjust decisions and laws, we only perpetuate their evil. By protesting an unjust decision, we are only voicing our rights.

Your Honor, I understand your position in this courtroom and your need to follow what you consider to be law, but what I participated in was civil disobedience and I am now ready to suffer the consequences. I am not

By our strength of feminism, consensus democracy and nonviolence we can do more than save the trees. We may be planting the seeds of a new civilization.

speaking to sentence to make an apology for my actions; instead, I wish to state for the record that I am proud of the stand I took and I am proud to once again feel the strength and hope which reverberated around the work this summer.

Editors' Note: Ms. Savile went directly to jail at the end of the trial.

◆

PETER SCOTT: ...I do not hate loggers. I do not hate judges. I have no contempt for you or for any of the people in this court....

I have seen people cry in this courtroom. I have seen a pregnant woman stand in a different courtroom and ask for a leave because of the pain and I saw the court had to weigh this. It upset me that this had to be argued, that her actual appearance would not be enough to allow her to leave. I submitted myself to the judicial process not out of contempt for the law but of respect for it....

The tide is shifting, Your Honor, and I wonder if institutions will crack under the weight of it. I am a Canadian, Your Honor, and a proud one, but now I have to make a distinction between the people, the nation of Canada and its state.

With one statement by yourself, you have made me an enemy of my country and perhaps stated a conflict; the people here and the state there. I hope that will not be so.

◆

DAVIDA SEFERITH: ...As I stood on the road, I really tried to conduct myself, Your Honor, with peace in accordance with the principles put forward by the Friends of Clayoquot. I conducted myself in every way possible to make it apparent that I was not a man of violence and that I would not promote violence....

[The following, concerning Charter rights, is from] Chief Justice Dickson from Regina vs. Big M. Drug Mart. This statement has to do with the concept of life and the security of the person:

> "...It would be incongruous if a Charter which implied no less than the Canadian Bill of Rights R.S.C. 1970, expressed in its preamble the dignity and worth of the human person were to shield a person from the loss of a finger, but not from the loss of her self-respect."

...It feels that this court has attacked my person in many ways. It has certainly attacked my intellectual integrity and it seems my self-respect. It has made implications that my intentions were criminal. I guess when people feel powerless,...it's a real struggle to maintain self-respect.

◆

TERESA SHANKS: ...I admit I am in contempt not of this court but of MacMillan Bloedel's logging practices. I was arrested because I refused to sit idly by and watch the trucks roll....

I feel honored to be one of such a wonderfully diverse, determined and strong group of people. As far as sentencing goes, I am a full-time student at UBC and also work at two part-time jobs. I work with community home supply services, which entails working with the elderly and disabled in their home. This trial has interfered with both school and work.

I have had to commute back and forth from Vancouver, which has been financially and emotionally taxing. I have missed a considerable number of classes and I have had to withdraw from one course. I have no money....

◆

JANINE TOMANEY: First of all, I do not have any money to pay a fine. If I have to go to jail, I will probably lose the apartment that I just got and I will not have anywhere to go when I come out, but I still think what I did was right. I do not think I did anything wrong.

I am only 18 years old but...I can see what is going on in the world. I cannot swim in the lakes around my house anymore.... I am so scared that when my son is 18 he will not know what it's like to sit under the sun because it will kill him. I think what is going on in the rainforest is horrible. I had to stand on the road that day. This world is just going to die if we don't do something soon. We cannot live in glass domes.

◆

JOHN VEDOVA: ...When this trial began I believed I was naive, but after many weeks to me the law seems naive. The law acts like a tempestuous three-year-old feeling like the center of the universe around which everything revolves. It has covered its ears defiantly so as not to deal with the unpleasantness of truth and justice. Like a three-year-old, the law wants its own way. It only wants to eat dessert but here we are the brussels sprouts,

attempting to bring back balance to a legal system that has become constipated and cranky when faced with healthy dissent.

◆

JAMES MILLAR: [I am a] lawyer speaking for Ms. Kato [who] is a market research surveyor, self-employed for the past 15 years. She always obeyed the law, respected the law. No prior criminal convictions, no prior breaches or injunctions. She heard about the issue and wanted to see Clayoquot so she joined her friend from the church choir in Vancouver, Inessa Ormond. She did not go to be involved in a blockade, she did not contact the media, she did not speak to the media, she signed the recognizance, she believed it was a civil contempt and she walked peacefully to the paddywagon. She has not disrupted these proceedings at all. She advised that had she known her actions would lead to a conviction for criminal contempt, she would never have joined the blockade.

Her intentions were to protest forestry practices and preserve the rainforest, not to bring the court or its order into disrepute. She says that all that is necessary for "evil to prevail is for good persons to do nothing."

You do not look at the amount of media coverage, you look at the intention in sentencing and Ms. Kato's intention with respect to blockading was directed towards forestry practices and she was negligent with respect to defiance of the court order. It is a major mitigating circumstance, in my respectful submission, that the actions of Ms. Kato and others were peaceful.... Sergeant Doyle told me yesterday, "My hat's off to the Friends of Clayoquot for keeping this so peaceful." He said that he personally had a hoot on seniors day. It was a lot of fun. I think as well you have to look at the government action—their recognition of the validity of Ms. Kato's cause and these people's cause and the recommendations of the Owen Commission that would have seen a moratorium on logging in the Clayoquot.

I stress that people ultimately be discharged, for surely they are not criminals. Ms. Kato travels internationally often. She should not be branded a criminal political subversive when she goes into another country. There should be a discharge in Ms. Kato's case because that will amply demonstrate the resiliency of the courts.

◆

Our silent protest was directed at the government for allowing the blatant destruction of our last remaining temperate rainforest on Vancouver Island.

LAWYER, JOHN OMAN: Faith Moosang and Liz Meyer are each 27 years of age. They were concerned about the forests and they were there to be counted. They have not disrupted these proceedings.

Ms. Meyer is registered at the West Coast College of Massage Therapy, involved in intensive study. She also works part-time. She has agreed to participate for five hours of community outreach at a hospital dealing with individuals who have nerve problems and assisting them as a form of volunteer work.

Ms. Moosang is enrolled full-time at Emily Carr College of Art and Design studying photography. She has an honors Bachelor of Arts degree and spends at least 50 hours a week full-time at school and works 10 hours as a research assistant.

Ms. Moosang and Meyer as well have co-edited a book entitled, *Living with the Land: Communities Restoring The Earth*. It is used in studies throughout North America. Incarceration would certainly disrupt their lives. I would recommend a suspended sentence.

◆

Clayoquot physician Ronald Aspinall, M.D., prepared this sentence plea:

I have spent approximately six-and-a-half weeks in the trial process. In this period I have personally spent 28 days in court, the majority of these days spent hearing the defence of others and not in regards of my own defence. I have spent two days here in Victoria waiting for adjournments. I have spent the better part of 13 days travelling to and from my home in Tofino, and more recently from my home in Ahousat, to court in Victoria.

Further, I have already spent four days in Nanaimo Correctional Centre following my arrest (for refusing to sign an unconstitutional release order precluding my rights to demonstrate at the Kennedy River Bridge and not knowing for which incident or incidents I was being charged). Because I have already incurred significant penalty in the trial and the trial period, I deem 47 days compensation in the assessment of the sentence.

Having only been married on May 29th, 1993, seven weeks prior to my arrest, I have spent as much time in this trial as I have been married.... The separation I have already had from my wife during my jailing and trial has

been a considerable strain on both of us....

While it may seem of relatively small concern in a society that tends to spend more time on acquisitions than relationships, my wife and I have lost the opportunity to spend the month of September canoe-camping in the Sound. September is the most beautiful month for this and the salmon are going up the streams....

Uncompensated injuries—criminal: I have received uncompensated injuries from MacMillan Bloedel (MB), or Ministry of Forests (MoF), or forest workers (FW), for which I have not been compensated:

(i) No compensation from MB for destroyed bicycle 1991.

(ii) Not able to proceed against FW for murder threat 1991.

(iii) No charges laid against MB for hit/run accident July 6, 1993.

(iv) Not able to proceed against FW for vandalism of car July 9, 1993.

I have spent an estimated $1,000 in travel, food, telephone, subpoena and billeting expenses during my five-and-a-half weeks in trial.

Potential Censure by College of Physicians and Surgeons...if there is a finding of guilt of criminal contempt.

Potential denial of Australian Immigration...parents and family live in Australia. MacMillan Bloedel has received relatively minuscule fines. The persons responsible never faced jail. In 1988 a forest protector was shot three times by a Fletcher Challenge employee; the case went to diversion, and the forest worker was given 20 hours of community work as penalty.

◆

Some protestors wrote out their views as did Ms. V.R. Burgoon:

That my name is Vanessa Ruth Burgoon and I was on the road at Kennedy Lake Division on August 9th, 1993.... That my stand on that day was not a stand to dishonor the courts....

That there are a myriad of reasons why I chose the stand of a conscientious objector.

That I am a National Lifeguard and I chose this field of work because of my respect for water and life....

My father worked as a cedar salvage operator in the Kennedy Lake Division.

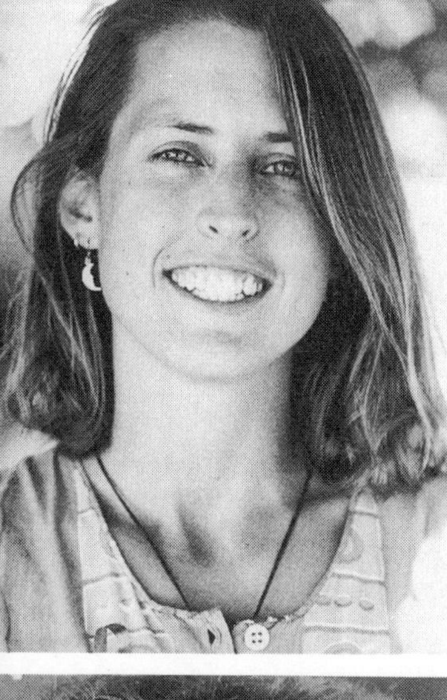

Remembrance of this destroyed land and its resultant dangerous flooding made me step forward on the 9th of August to prevent this same destruction occurring to the remaining temperate rainforest within the Clayoquot Sound region....

That I have visited archaeological sites and know that this land was and is inhabited by the original peoples.

That on August 9th no land settlement with aboriginal people of the Clayoquot region had been negotiated and therefore the jurisdiction of this court is still in question and that due to this lack of transaction at any time in history the courts do not have jurisdiction over the road at Kennedy Lake and that I therefore did not consider myself in contempt but that I consider MacMillan Bloedel to be in trespass on said land....

I chose to support our country's agreement made at the Earth Summit.

RONALD MACISAAC: May it please Your Lordship. "The best and the brightest will be at the blockades this summer," said famous lawyer Clayton Ruby. "The world is laughing at us. We are a civilized world power jailing our intellectuals, for trying to save a planet hell-bent on self-destruction...."

How does Canada punish those who do not honor the directives of the court? Well, when much-respected and revered Chief Justice Wilson told the Social Credit Attorney General of British Columbia to retry the forest companies accused of bribery, his court order was not obeyed.... This Attorney General went later to MacMillan Bloedel as its vice-president.

Then there are those most dangerous of all criminals, the drunken drivers. When they are ordered by the court to cease and desist from driving and they continue to drive, they are charged under the code and given small fines. But the people who protest the destruction of the rainforest and thereby disobey the court order, is their sin as bad as the driver suspended for his life-threatening disobedience, or the Attorney General who ignores the directives of the Chief Justice and does not pursue the sins of the forest companies and its leaders?

With regards to the severity of the offence, the injunctions were based on a SLAPP suit. This is now illegal in Washington and Oregon and I suspect will soon be illegal in the province of British Columbia.... Of the first

Having only been married on May 29th, 1993, seven weeks prior to my arrest, I have spent as much time in this trial as I have been married....

seven hundred protestors, their education was found to be largely post-secondary and their unemployment level was only four percent. In other words, it is the establishment against the establishment. With regards to policing costs, in the case where MacMillan Bloedel was recently fined $35,000 dollars for breach of logging regulations they were allowed to keep logs worth more than thirty-five thousand dollars....

I was optimistic enough to think that all of the efforts that have been made from time to time during the course of this trial on a political level would result in a consensus and a healing and a stopping of what I would consider devouring our own young. And I file with the court, rather than reading it out, that our Attorney General in British Columbia is considering anti-SLAPP suit law....

◆

THE COURT: Tell me something, Mr. Parker. I read in the *Globe and Mail* this morning that you were reported to have said this trial was a travesty of justice. Did you say that?

STUART PARKER: Yes My Lord, I did.

THE COURT: Thank you.

STUART PARKER: ...[In] 1992 I was the chair of a youth forum organized by the B.C. Black Educators. My family is from a long line of black Canadians who have been involved as politicians and athletes representing Canada internationally in different capacities. At the conference one person asked the question of the panelists, "What is it that defines us as black Canadians, what is it that we have in common...?" The answer was it is the struggle that defines us.

I come from a line of people...who until the 1960s in this province were not afforded the basic protection of the law, who until the 1960s were not allowed to vote....

It is always said on paper in the United States that all men are created equal, that all people are created equal, but it is through civil disobedience, through the resistance of unjust laws that the constitutional rights of black people of this continent have been won. My grandfather in the 1950s was a porter on the CN Rail.... As a black person he could have no other job with CN Rail. Vancouver weekends he spent his time sitting in cafes trying to be

served a meal, a basic right on paper extended to everyone,...and he was taken away by the police for trespassing on the premises of these restaurants.

It is all very well for the rich, white men who have always had the protection of the law to talk about descent to anarchy, how the rule of law, how the rule of justice will be destroyed if we resist the will of the courts or the will of governments through civil disobedience. But it is through civil disobedience that ninety percent of the people in this country have won the democratic rights that we now cherish.

It is my belief, My Lord, that civil disobedience has done more for the cause of justice in North America than obedience has in the past four hundred years and I am proud to say that I continue to be part of the struggle for the rights of those that are not enshrined yet in this constitution and in the rule of law—the trees, the animals, the insects, all of the lifeforms that live in our dwindling ancient forests on Vancouver Island and throughout this province. I am proud to say that I will continue the struggle through legal and, if it continues to be necessary, other than legal means to see that the cause of justice—as in the U.S. Civil Rights Movement in the 1950s, as in the Women's Suffrage Movement at the turn of the century—that the cause of justice continues to triumph over the law. Thank you, My Lord.

◆

ROBERT MOORE-STEWART: First I point out that the term political speech is vague, indefinite, undefined. What I am concerning myself with is speaking the truth as I understand it.

THE COURT: You have fifteen minutes.

ROBERT MOORE-STEWART: ...Sentencing, Your Lordship, is supposed to be a time of broad focus in taking in the overall context, the whole context surrounding the events the accused is charged with. It is a time to take off some of the restrictions and many of the constraints, rules, and rulings of relevancy that otherwise constrict the range of focus for the court....

I wish at the beginning to refer to certain pages of Your Lordship's decision, written judgment of yesterday, not to reargue conviction or acquittal but to make comments or correct errors that may affect sentencing....

"The will of the people as fortified by a court order is held up to ridicule and mockery...."

"The world is laughing at us. We are a civilized world power jailing our intellectuals, for trying to save a planet hell-bent on self-destruction...."

The will of the people...? It's got to be in that context that the will of the people is supposed to be the continuation of the clearcutting of the old-growth forest in Clayoquot Sound. That is not the will of the people....

Ridicule and mockery of the courts was not what was going on on the logging road at Clayoquot Sound. The court is seeing insult and mockery where none existed or was intended. Following the arrest for contempt, the process has...criminalized the otherwise civil complaint of MacMillan Bloedel....

Page 6 lines 17 to 25 of your judgment, I will read the paragraph:

> "Democracy and the rule of law allow every person to protest by peaceful means. But once protestors infringe upon the rights of the others, the law must step in. It does so for two purposes. First to protect the legal right and second to prevent violence. If it did not, then the only way of settling the dispute would be by force of arms. In this case, it would be the forest industry and its allies versus the defendants. There is little doubt who would suffer the most."

First I want to talk about the concerns of the law in regard to protecting legal rights of persons, and I suggest that in regard to the way the whole administration sometimes functions, some persons may be more equal than others. I suggest that multinational forest giants such as MacMillan Bloedel are trashing our old-growth forest, ruining the soil forever, desecrating the fish and salmon habitat, and destroying the forest. All that is against the law. To begin with and not to end with Section 33 of the Fisheries Act, making the destruction of fisheries streams an illegal, indictable offence—

THE COURT: Okay, you have five minutes.

ROBERT MOORE-STEWART: I object. I will continue, under objection, thank you. All this you must set in the context on sentencing, I submit. If the legal rights of MacMillan Bloedel are being enforced by the judge...then the courts should in all fairness, I submit,...be aware...those alleged legal rights were founded on fraud....

The second purpose of the law mentioned is to prevent violence.... The severe and only violence that is taking place is that of...MacMillan Bloedel['s]:... destruction of the old-growth forest, and all that lives within it,...starving the animals....

It is destroyers of our life-support systems that are threatening us with social breakdown, just as in Russia, Somalia and Lebanon—

THE COURT: You can tell that to the Court of Appeal. Everything you say now you can tell to the Court of Appeal. Do not tell me. Get on with the sentencing.

ROBERT MOORE-STEWART: ...Any harsh sentencing of these defendants would likely increase any unfortunate perception of the courts as being biased for the multinational logging company and against the people, the animals and the forests in supporting this status quo of destruction—

THE COURT: One more minute.

◆

DAVIDA SEFERITH: ...I would like to make a submission on your ruling because you promised we would have full chance to give arguments. Much of the arguments, the prosecution is filibustering and you are preventing us...from giving our defence.

THE COURT: That's my ruling, you have to live with it. The Court of Appeal can maybe correct me. I am giving you fifteen minutes each.

◆

THOMAS BELLAIRE: ...I see now that the courtroom is not the place to settle our dispute but I have not chosen this medium, the court did when it chose to get involved in the forest....

We have bashed each other back and forth for six weeks; no one has given in. What we have in the end is not justice but one side using power over the other to end the dispute. This may hold the problem at bay for a time....

◆

TERRY BROWN: Since I [was] a child my first love has been for the Earth, for exploring the Earth. Relating to it has led to expression in my photography and my writing. It is very much part of the core of my being, who I am—it defines me. Along with that has been my development of spirituality.

I believe in a God who is a supreme being, the great spirit, the creator. My background is in the Christian Church. I am a former clergyman...and although I am no longer a part of any organized church I still believe in God...that's the central integrating source of my life.

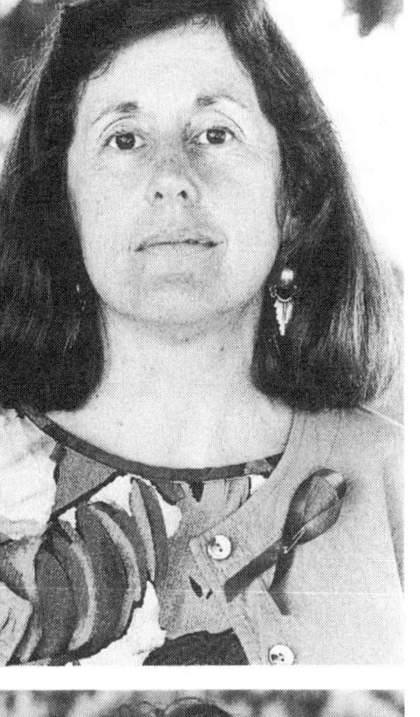

I believe as Psalm 24.1 says in the Old Testament scriptures, "The Earth is all the Lord's and everything in it, the world and all who live in it." For me this is the basis of justice, that the creator is the one who ultimately owns the Earth; we as human beings do not.... The creator is the rightful owner of this Earth and that we are stewards to take care of what the creator has given to us. That caring should be rooted in love. A passage from Dostoyevsky's writings, *The Brothers Karamazoff*, reads like this:

> "Love all of God's creation, the whole and every grain of sand in it. Love every leaf, every ray of God's light, love the animals, love the plants. Love everything. If you love everything you will perceive the divine mystery in things. Once you perceive it you will begin to comprehend it better every day and you will come at last to love the whole world with an all-embracing love."

I submit to you that my actions were motivated by this love. I would like to read a portion from a journal that I wrote in 1987 on a second visit to Clayoquot Sound:

> "As we drove through the dusk on Thursday evening I could see evidence of clearcut[ting] right down to the highway in places. I was appalled since I don't remember seeing any of that when I was at Long Beach in 1981. Have you ever heard a mountain cry? Well I did that evening as I saw a mountain that was completely logged, clear-cut to the bone, the ancient bedrock.
>
> "I could hear it crying out to my spirit, a heart-wrenching weeping as it stood there in disgrace, ashamed of its total nakedness. Shorn of its glorious attire of trees, moss and animals it stood exposed to the gaze of all who passed by. The mountain cried because of the wounds caused by the men who raped her but also she cried for the souls of that race of beings who committed this crime.
>
> "The mountain sought for a member of that race who could hear her cry and could understand in some small way what she was saying. My passing that evening was no chance encounter. I can still hear the mountain speaking to me and I pray that God will give me the sensitivity to learn from what the mountain says and the wisdom to know how to translate the message into a form other humans can understand.
>
> "It seems to me that among God's created works only a mountain has lived long enough to know the true value of a tree. We humans in

"To suggest that these two boys and thousands of other young people across the province do not understand the issue is patronizing and utter nonsense."

our arrogance think we know the worth of a forest but we live much too briefly to understand the value of the forest over a thousand or ten thousand years. Our perspective of worth is too short and much too pragmatic in a self-serving sense...."

I record a prayer that I prayed in 1987 in July:

> "God, I pray that I at least will be a willing listener to you and to your creation so I will know the value of the Earth and its various ecosystems and I pray that I will be a spokesperson for those created things which are not commonly heard by people. Father, may your kingdom come to this Earth through this child and learner. Amen."

And I read that journal entry a few weeks ago, after I was arrested—I had forgotten about it completely. What I said when I was arrested and taken away was....

> "Here I stand for the wild things that have no voice in our court, in our boardrooms and in our politics. I speak for the wolves, the trees, the eagles, ancient forests forever."

And when I read this prayer that I had prayed and written six years ago, I heard an echo of what I had prayed and said, then I saw the completion of that circle—that I was finally being able to speak for the wild things that have no other voice in our society and for me it was a very moving and profound experience. I understand your role in this is just the legal aspect, but for me it goes far beyond that. I stood on that logging road and I stand here in court facing the consequences because of the love I have in my heart for all my relations. Thank you.

◆

GEORGE HARRIS: ...I have spent five hundred and eighty-two dollars on ferry, food and gas over the last five weeks and my lost wages on my sawmill are approximately $4,000. I have lost two future contracts because of this, so basically my year financially has been ruined by this court case, which will impact considerably on my ability to pay the fines which may or may not follow....

My son Tyson Harris, age eleven, and my son Adam, age twelve, feel quite responsible for the situation that I am in. They both asked if they could serve the time alongside me in jail....

These kids are taught in school to think critically and to take responsi-

bility. They have learned on Galiano Island beyond a reasonable doubt that what MacMillan Bloedel is doing is immoral and illegal. I had hoped to prove...to Your Honor that MacMillan Bloedel has been breaking the law and is continuing to break the law. And I feel that my job here is as spokesman for tomorrow's generation. Adam and Tyson know there won't be the forests left when they grow up.

They know their father makes a living in the forest industry and they think that what we are doing in this society is immoral and illegal. I would not have gone to Clayoquot Sound on my own accord. I am not an environmentalist as such and don't belong in any organizations. I have never met any of these people in this courtroom before, although now I would have to say I am extremely proud to know every one of them; they are an amazing group of individuals, Your Honor.

THE COURT: Let me ask you something, Mr. Harris. You are a logger.

GEORGE HARRIS: Yes sir.

THE COURT: Suppose some people had come along and sat down in front of your roads and had not allowed you to get your logs out so you could make a living. What would you do?

GEORGE HARRIS: ...Well if I could not convince my neighbors and my community that what I was doing was responsible, I would quit what I was doing.... I wanted to quote from the *Vancouver Sun* editorial page on July 10th:

> "The recent arrests of Tyson and Adam Harris on the logging roads of the Clayoquot Sound...captured in a single event, the despair and urgency of an entire generation. Premier Mike Harcourt's response—'I think it's really unfortunate and I am disturbed that children are being used that way once again'—shows that he has really no understanding of the concerns of young people. Rather he sees Tyson and Adam as merely pawns in an elaborate chess game between adults competing for money and posturing for the division of natural resources.
>
> "To suggest that these two boys and thousands of other young people across the province do not understand the issue is patronizing and utter nonsense. I am sure both Tyson and Adam feel the destruction of the chainsaw every time they play out on the east slopes of Galiano Island where MacMillan Bloedel has already left its legacy. Who is supposed to ensure that Tyson and Adam and their children

It is my belief, My Lord, that civil disobedience has done more for the cause of justice in North America than obedience has in the past four hundred years

have forests to play in, admire, study and be healed by? The current government has let them all down by caving in to the forest industry's demands. By making decisions based on dollars and votes the government has once again breached its fiduciary obligations to future generations.

"Tyson and Adam are true heroes. With their tiny bodies and courageous spirits they stood up to the loggers and the state and let them know they *do* understand what Premier Harcourt will never understand, they care and know we are running out of time."

This is from the *Globe and Mail* editorial page:

"Indeed the prospect of logging something as nearly spiritual as a rainforest, one of the last of its size in North America, has become a soul-defining issue in British Columbia...."

I was thinking this morning, you know, what the parents of the eleven- and twelve-year-old children said in Nazi Germany in the 1940s when their Jewish neighbors were taken away to be exterminated. This was done legally, this was done with everyone's knowledge. What did the parents say? What will I say when my children grow up and look at the destruction of this province? I will say that I stood with my friends on the line at Clayoquot Sound. Thank you, Your Honor.

◆

GUY WERA: Needless to say I am not very ready to do this....

I never saw before how the court was undermining democracy....

If you give [a] few people the right to play with the law in any way they wish, and not allow anyone the right to criticize or to argue or to challenge this, I do not think it is a law, I do not think an injunction is law. Is an injunction law?

THE COURT: Yes.

GUY WERA: So law is made by very few people, which changes the behavior of a lot of people and if we are not allowed to challenge that—

THE COURT: You are. You are allowed to challenge it. Appeal it to the Court of Appeal.

GUY WERA: But before we go to the Court of Appeal we have to be here.

THE COURT: You can go to the Court of Appeal. The injunction which was issued and the court acted upon against you was issued in 1992. If you did

not like it, you could have gone to the Court of Appeal.

GUY WERA: I did not know of its existence until now....

◆

MARGARET [INESSA] ORMOND: I did not go to the Kennedy Bridge as an environmentalist. I had never given myself that honorable title although I recycle and use environment-friendly products and do not pollute the atmosphere with an automobile. I went to Kennedy Bridge, not as an environmentalist—I am just very emotional this morning—not as an environmentalist but as a private citizen, out of concern for the world my children and grandchildren will inherit.

I was worried years ago when I heard about rainforests being chopped down in Brazil, and the effect that would have on the environment. I was aware that clearcutting also occurs in B.C. but I felt helpless, thinking that the forest companies doing the clearcutting owned the land on which they were working.

I was shocked this spring to learn that the forest companies are given licences by the provincial government to clearcut Crown land, public land and land that is subject to native land claims.

At that point I felt I had a responsibility as a citizen to stop the logging companies from doing irreparable harm to land that we need to return to the First Nations in the same condition as when it was taken, and to stop the logging companies from clearcutting the rainforest that provides cooling and oxygen for all the Earth's creatures.

The Crown says we had other legal alternatives than blockading. I was at a disadvantage. I did not know what they were.

MacMillan Bloedel has well-paid corporate lawyers who know how to use the courts to their advantage. I did not know what law prohibited us from protecting the forest or how it could be challenged.

I do not know if there is sufficient time to legally challenge the law before irreparable harm is done.

I think irreparable harm has been done every day this summer that the logging trucks have passed Kennedy Bridge. MacMillan Bloedel flagrantly breaks the law daily in Clayoquot Sound and I do not know how to take them to court for their illegal logging practices. I do not have the money to

hire lawyers to do this for me.

I went to Clayoquot Sound to protect the forest from the logging companies. I stood on the Kennedy Bridge July 5th and refused to move even when the first MacMillan Bloedel truck came within six inches of my toes. I was scared. I remembered the movie, Gandhi, and the army firing on the people. But I held on tight to my fellow blockaders and stood strong. And the truck backed away and left. I stayed all day by the bridge, just in case they might come back. I went back to the bridge just in case they might come back. And I went back to the bridge the next day and blocked the logging truck until I was arrested. I would have stood there all day and every day if I had not been arrested.

I could not bring myself to sign an undertaking not to go back to the Kennedy Bridge and stop the logging trucks again, so I stayed in jails and prisons for two months until the first day of this trial. At that time I signed a promise to appear in court, because it had always been my intention to appear in court. My contempt was not of the court, but of the logging practices of MacMillan Bloedel.

I want to share with you some of what I learned in prison. I learned that to be enclosed by walls and kept from contact with the Earth kills the soul a little every day.

And I know that not only the people in prison are dying in that way, but also the people in cities. And through the deprivation of freedom and free contact with the Earth I came to a deeper love and passion for freedom and the Earth.

If there are indeed three arms of government and if they all act to support big business and its greed, which weakens and destroys the Earth—our life-support system—where are people to turn to for justice? The MacMillan Bloedel court injunction indicated that there is no separation between big business and all three arms of government. This court, in continuing to uphold the MacMillan Bloedel court injunction, will confirm that this is so.

We might forgive past generations for their greedy ignorance, but there is no longer an excuse for ignorant greed....

There is no excuse for the continuance of damage to the environment. It is time to change the way we think, to refuse to accept substitutes for our

genuine needs, and to start respecting the Earth, which provides us with what we really need. And this change needs to be reflected in our laws and court orders for justice to be served.

I had hoped to have expert witnesses here to testify to the unjust MacMillan Bloedel court injunction that I have challenged. The court has not allowed it. If the court sees as its duty to uphold only the letter of the law, rather than justice, it is itself welcoming contempt as well as in the long run threatening democracy and the survival of the human race, because we will cease to have a life-sustaining, soul-sustaining planet on which to live.

I urge you, Judge Bouck, to go up to the Clayoquot, to walk on the Witness Trail and then continue to the Black Hole. See if you are not moved by the beauty and by the devastation.

◆

KENNETH HOWES: ...The reason why this trial is so irregular is because it has been significantly designed and influenced by the Attorney General of British Columbia to give utmost advantage to the plaintiff. The nature of the charges and the latitude given the court in pressing criminal charges without being restricted to the criminal style of judicial process, the denial of trial by jury—all these factors work to the detriment of the accused. To have the mechanical requirements of due process is insufficient if the arguments are not fully made, and if there is no jury to hear what is at stake and to make a decision on the justice of the defenders' action and not solely on its legality. As the distinguished 20th century legal scholar Wigmore wrote: "Law and justice are from time to time inevitably in conflict...the jury and the secrecy of the jury room are the indispensable elements in popular justice."

We are drifting fast towards the 21st century and we need to develop new and creative approaches to the problems of our age. For citizens to think for themselves about the great issues before mankind is to witness democracy at its purest, it is to see democracy come alive.

These are not the times for the courts to be breeding apathy, for like a strong wind the rulings of the court are powerful and far-reaching and they will carry off all they are able to lift. Including freedom. Including justice. Including our ability to improve our world.

◆

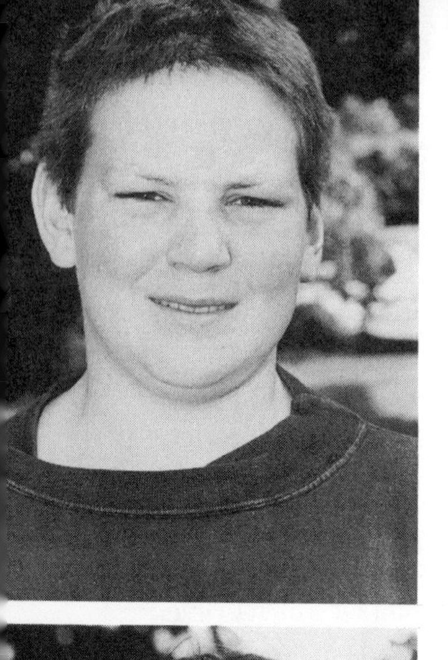

The next speaker was one of the grandmothers well-known for taking jail before freedom rather than signing a commitment to stay away from MacMillan Bloedel.

BETTY KRAWCZYK: I have read the reasons for your judgment several different times and although you have stressed your points of view in this matter repeatedly during the course of this trial it was instructive for me to read your comprehensive analysis of why you have ruled the way that you have. And there is one thing that I can say for this document sir, it is awfully consistent. You do not deviate from your original thesis that nothing is to be considered in this trial except whether we, the Clayoquot protectors, disobeyed the injunction order.

From this narrow definition, without considering motive, you find us all guilty of criminal contempt. You say you have had to hold to this narrow definition of the law because to consider any other would be to allow feelings or emotional involvement with the Earth to interfere with arguments that are strictly legal. But I submit, sir, that while your arguments are couched in legal language they also show emotional bias. I refer to the top of page 13 of your Reasons for Judgment: They argue that they were merely—

THE COURT: Just a minute ma'am. Let me say this. This is not the place to point out where I may or may not be right in that judgment. The place to do that is in the Court of Appeal. This is the time to assume that what I said was right. You might not like it but assume that is the case and tell me what you would like to tell me with respect to sentence.

BETTY KRAWCZYK: That is very harsh, sir. Truly it is. I thought this was to be a sort of overall presentation of why we feel we should be sentenced in a certain way. Is this—?

THE COURT: Well, that is right but you seem to want to get in and start criticizing the judgment.

BETTY KRAWCZYK: ...The only thing I wanted to present, sir, is that you say you cannot listen to emotional arguments.

THE COURT: I do not think I said that. I said I have to decide the case according to the law. That is my sworn duty. That is what I said.

BETTY KRAWCZYK: Well, I would just like to make this one point about your argument and that is all, just this one point, if I may. You say that: "In

I went to Kennedy Bridge, not as an environmentalist... but as a private citizen, out of concern for the world my children and grandchildren will inherit.

my view, the logging protest defence was simply a form of moral excuse they used in order to try and justify their unlawful behavior."

I find this sentence very interesting because you seem to be suggesting that our unlawful behavior was inherent—that is, all of us were and still are, I suppose, possessed of a primary and free-floating state of unlawfulness that needed only an object to fasten upon like the clearcutting in Clayoquot Sound, in order to reveal itself to this court as poisonous and antisocial.

Are you really suggesting, sir, that if I were not in that prisoner's box for blockading a MacMillan Bloedel logging truck, that I would be here for some other unlawful activity? Are you saying that my particular logging protest defence is an excuse to try and justify any otherwise inherently unlawful behavior?

I cannot remember whether it was the Vikings or the Greek people of Sparta of whom it was said that when the warriors went off to do battle and were killed in battle they were ceremoniously brought home on their shields.

The women who sent these warriors off to do battle told them either come home with their shields or on them. Well I promised those poor bleeding mountains behind my house at Cypress Bay that are still bleeding from those landslides from the clearcuts of ten years ago, that I would come home with my shield or on it, and I intend to do that, sir.

I have learned a lot of things these three months I have spent in correctional centers. I have always been claustrophobic. I do not like to be closed in. I do not like the closed-in space of cities or small houses or small apartments. I like the open expanses of the sea and the mountains and forest, and the very concept of being locked in behind bars in some closed-in, cramped space has always filled me with absolute terror.

I tell you, sir, if I were not fighting for my own home and the rights of my grandchildren to that home, in addition to simply fighting to preserve the forest, I would never have been able to refuse to sign that undertaking knowing that these worst nightmares of mine were right on the horizon. However, I found, as people often do who entertain the very worst scenarios in any given situation, my fears were worse than reality. I can be locked

up now without batting an eye. I can even eat institutionalized food.... As the song we sing goes: I truly am stronger than I was before.

Well, in addition I came from a very large family and I have been the single matriarchal head of my family for the last twenty years and I think one of the reasons I have had such a difficult time dealing with you, sir, is that in my own clan I am the judge and my children and grandchildren look to me to settle disputes and to give advice and to help determine what is right and what is wrong, and I have always taken this role seriously.

This started as soon as I became aware as a young matron that the law of the land could not only be wrong but downright evil. I was living in Baton Rouge, Louisiana, with my husband and children when the integration order of the Supreme Court came down in Louisiana. New Orleans was the worst of the disturbances about...the integration of the schools, but in Baton Rouge the elementary schools that my children attended closed rather than integrate.

My consciousness was suddenly raised by leaps and bounds. For the first time in my life I was forced to take sides on this issue and I really appreciated what my young friends had to say this morning about the racial issues and all of the emotions and contradictions that this issue always brings forth. In this instance, it took a federal law to override the state law and it was the civil disturbances that brought all of this to a head.

I went down and joined the civil rights groups and we picketed the school to open and integrate, but I was definitely a Johnny-come-lately. I felt ashamed that I had to wait for a federal order to do what I knew was right to begin with and I vowed after that to never, never hold still while laws protect the very worst kinds of injustice.

But the next test was not very many years over the horizon. We had moved to Virginia and my husband was now a physicist and went to work for NASA and in 1963 my oldest son left the university and joined the Air Force—and I hope I do not forget what I want to say because that is all I had time to write—but I did not know anything about the war in Vietnam. You know, I was very busy with the affairs of a women with a big house and a lot of kids and a lot of different types of activities. But when my son joined the Air Force I thought I had better find out something about the

Vietnamese War.

The more I read, the more horrified I became about how long these Vietnamese people had been fighting.... The United States and some other countries had arbitrarily divided Vietnam. The Vietnamese people had been fighting for a hundred years and had just finished defeating the French and now were having to start all over with the Americans.

So I became extremely opposed to the war. So did my husband and so did our next two sons who were in line for the draft. My husband was told by NASA that if he continued to protest the war he would lose his security clearance, and that really—again my consciousness took a radical leap because here we are in a situation that says you are going to lose your job, this wonderful job you have got, this house, your life, your family, unless you stop protesting what the law is doing. And that, sir, is how we all wound up in Canada in the first place.

But after that there came another situation which in a way was even worse: the feminist movement burst into the consciousness of society and I found that all of these things that I had been struggling with all of my life in one form or another, all had a common cause, and this cause is that we live in a patriarchal structure. We live in a capitalistic structure, we live in a land where more laws have to do with property than have to do with people.

Very often the laws are written and designed primarily to protect property in the sense of privileged property, corporations, companies and there is just a lot of injustice around and if people do not start becoming in a sense their own judges and in a sense again [having] some concept of being able to decide for themselves what is right and wrong within the context of the laws—this is not anarchy, it is becoming adult, becoming a responsible citizen. Thank you, sir.

THE COURT: Thank you, ma'am. Mr. Kubiki?

GRZEGORZ KUBIKI: ...Despite Your Lordship's guilty verdict in my case I do not feel guilty of any contempt of court. I do not consider myself a criminal and I doubt if anyone who knows me will ever be prone to consider it, and I have letters from my professors for Your Lordship.

I am not afraid of having a criminal record. Just a few years ago thousands of Poles were given criminal records for the struggle to defeat the

I can be locked up now without batting an eye. I can even eat institutionalized food.... As the song we sing goes: I truly am stronger than I was before.

totalitarian monster.... I came to Canada because democratic ideals seemed to me very attractive, not because it is a land of opportunity as some say. In other words I love Canada for what she offers, not for how much I can rip her off. This is, I think, the reason why I found myself in this courtroom.

What I find extremely sad, the only money I expect to have is for my education. I have no income but a student loan and it would be against the law if I spent it for other purposes than educational. Moreover because of my involvement in this issue and the extraordinary length of this trial I suffered enough in terms of my academic work and private life. I can also expect other problems from the immigration authorities for my devotion to Canada's cause, what I could not foresee before....

THE COURT: Thank you sir.

DARRIN MORTSON: Your Lordship, I would ask that were I a person of foreign tongue, unfamiliar with English, standing accused before this court would I receive a translator? If I were a person deaf and [mute] standing before this court, would I not be able to communicate my defence in sign? I submit therefore, Your Lordship, that I am as unfamiliar and handicapped in this court as the foreigner and the [mute]. I submit that I am legally illiterate. Legal terms and language games cannot possibly for me express my defence....

It was a spiritual, post-linguistic act between myself, the trees, and my other brothers and sisters in the forest. To express this in legal terms would be to transform my passionate feelings into mere words. Banal, diluted and fundamentally untrue.

As I am legally entitled to defend myself I would ask that I would be able to do so in such a way that stays true to my feelings. If I am not allowed to do so I will say that justice has died and offer no defence except my agreement with that already presented. That being said I shall now read from Dr. Seuss' *The Lorax*. (He then proceeded to read *The Lorax* in its entirety.)

◆

As the trial wound its weary way homeward, I set out seven appealable issues:

RONALD MACISAAC: ...Viscount Hewart, Lord Chief Justice of England, said in 1924: "A long line of cases shows that it is not merely of some im-

portance but it is of fundamental importance that justice should not only be done but should manifestly and undoubtedly be seen to be done...."

In the eyes of some protestors there was a perception of a mistrial stemming from various factors such as:

1. Perceived lack of time to prepare, exacerbated by;

2. Lack of willing lawyers to take cases;

3. Various objections made by protestors who did not identify themselves, [which] left behind a confused record;

4. Alleged restriction on prisoners as to paper and pen;

5. Agency representation not allowed in early stages of the trial, which is said to have impacted on preparation;

6. Requests for severance;

7. Refusal of jury trial.

Decisions binding upon this court indicated that there is no right to jury trial but it is submitted that one could have been granted so that, in the eyes of the observers of this very public trial, there would have been no perceived unfairness in the chosen forum....

PUBLIC AND PRESS RESPONSE

Justice John Bouck convicted the 45 and sentenced them to fines and imprisonment ranging up to 60 days in jail and a $3,000 fine. The penalties were the first and the harshest imposed by the courts so far.

—The Canadian Press, Vancouver

◆

These people would have received lesser sentences if they had been found guilty of assault, theft, drunk driving or even rape.

—M.G. Price, Ganges

◆

To brand as criminal many of our best and most conscientious idealists can only increase the distance between people of conscience and the state.

—Jim Andrews, Victoria

◆

These people would have received lesser sentences if they had been found guilty of assault, theft, drunk driving or even rape.

As they were being led away, the protestors asked why MacMillan Bloedel and other forest companies get off for their history of abuses to the Fisheries Act, while they go to jail? The question reminded me of the Left's old saw about the law being "an instrument of the ruling classes...." The lesson of Clayoquot Sound—economically, politically, legally—is simple, but harsh: we have inherited a structure of self-serving power that is on track, out of touch, and dipsey-doodling.

—Michael M'Gonigle, Professor, Environmental Management, Simon Fraser University.

◆

[For] recent illegal logging in Clayoquot Sound...the company was fined but got to keep the timber it cut, rather like letting a burglar keep the TV he's just stolen so that he can sell it later and offset the fine.

—Stephen Hume, *Vancouver Sun*, September 3, 1993

◆

Seeing the world through the eyes of her unborn child caused a Nanaimo woman to face prison by protesting in the Clayoquot. Now faced with a 45-day jail term, 23-year-old Leesa Heyward has to decide whether to serve the time before her baby is born, appeal and hope for leniency from the courts, or try to serve her sentence at home with an electronic monitoring device....

"A lot of the protestors were really dedicated people—doctors, ministers, even former loggers—but they just classified us all as a bunch of unemployed hippies," [said] Rosa Heyward.

—Lynn Wellburn, *Nanaimo Times*

◆

DEFENCE OF NECESSITY: How do you deal with a system that simply will not hear you? What do you do when your conscience finally will not allow you to stand by any more and watch the destruction of ecosystems? When will society at large be prepared to act to prevent the planet from being irreparably harmed by the insanity of clearcut logging? For some, these are very real questions.

I sat in the Nanaimo courthouse as Garth Lenz...tried to explain exactly why he had blockaded a bridge....

He explained that for him the blockade has been a last resort. With patience and respect, he pointed out that over the years he had spent countless hours going through the right channels, writing to the ministries, attending public meetings, giving public slide shows about the forest situation in the Clayoquot, and photographing this magnificent wilderness. His action on that day in defying a court injunction meant not contempt for the court, he continued, but contempt for the system. Knowing what he did of the regional effects of continued clearcut logging, as well as the planetary effects, his conscience left him no other choice than to disobey the law. As a consequence, Garth pleaded defence of necessity....

In the end, of course, it was disallowed....

As I sat in that chamber of hopelessly blinkered authority, I thought of the truly unusual efforts that had in fact gone into resolving the Clayoquot issue peaceably. And I thought of the many locals and others from further afield who felt cheated by the public process....

As one friend writes, "All five logging camps have gone ahead with precedent-setting levels of cutting through all the controversy and discussion. Three more previously untouched watersheds have been ravaged. At this time there are only two major rivers navigable by canoe left intact. Of the several smaller waterways, there are only eight or nine left, not much when you take the 36 watersheds that once supported the salmon...."

—Chris Plant, *The New Catalyst* No.25, Winter 1992/93

◆

At the end of the mass trials, environmental leader Tzeporah Berman faced charges of abetting the protest. She had been a powerful force in preventing reprisal by environmentalists against threats and violence. She was charged with criminal contempt of court for encouraging others to break the court order by her actions.

A MacMillan Bloedel representative [admitted] during the trial that the firm had lobbied for months to have the charismatic woman [Tzeporah Berman] arrested.... The injunction, renewed in 1993, prohibited only physically blockading the site.

B.C. Supreme Court Justice Richard Low dismissed the charge after four

days of evidence.

—**Kim Wested,** ***Victoria Times-Colonist***

◆

Tzeporah [Berman] presided over a small miracle at the Peace Camp: [she urged all to be] peaceful—and they were. Is there a judge who would not be proud to call her daughter?

Many a bitter environmentalist tore up her NDP membership card but...if a "conservative" like Gordon Campbell now flying the flag of convenience of liberalism ever takes the helm, it will be every tree for himself.

—**Anita Bundy, letter to the editor of** ***Monday Magazine***

◆

Dear Mr. Harcourt:

Writing this letter to you from the Burnaby Correctional Centre for Women to advise you of my disappointment at your Clayoquot Sound decision. It is a little like confronting a two-timing lover...how could you do this to me after my long and passionate involvement with your party...?

Clearcutting is a crime...a crime against life. There is no such thing as a little bit of clearcutting. That is like being a little bit pregnant.

—**Betty Krawczyk, Prisoner No. 03793924**

◆

Rarely has a government so consciously destroyed its traditional voting base.

—**Editorial,** ***Victoria Star***

◆

On September 15, 1993, Dr. Betty Kleiman, a pediatrician and an arrestee, and Joan Russow, a sessional lecturer in global issues at the University of Victoria, appeared before Justice Drake with an application to rescind the extension of the injunction No. C916306 MacMillan Bloedel Ltd. vs. Sheila Simpson et al.

Their attempt to rescind the injunction was based on the following grounds:

1. That the granting of the injunction could contribute to non-compliance with principles endorsed by Canada through international agreements, obligations that have been undertaken such as the Biodiversity Convention. They referred to a requirement to carry out an environmental assessment review of activities that could

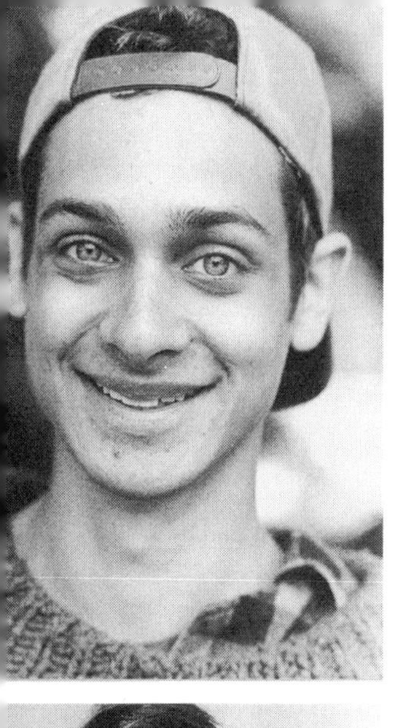

contribute to a reduction or loss of biodiversity. Practices typical of those carried out currently in B.C. forests have been assessed as being destructive of biodiversity. The practice of clearcutting followed by artificial reforestation leads to erosion, high irradiation and higher climatic extremes, [which] alter the microclimate,...soil compression,...eutrophication of groundwater, rivers and lakes....

2. The plaintiff must establish a right. The judge decided that MacMillan Bloedel has a property right. Russow and Kleiman pointed out that the right is a conditional [one] depending on fulfilment of statutory law, in particular the Forest Act and the Fisheries Act. Sections in the Forest Act and the Fisheries Act have been violated by MacMillan Bloedel. It was also pointed out in the application that the Ministry of Forests and the government of British Columbia have failed to seriously enforce Section 60 of the Forest Act [which provides the Forests Ministry with the authority to protect forests by cancelling a company's right to cut trees].

3.The third reason for calling for the rescinding of the injunction is based on the nature of the injunction as an equitable remedy moving with time and circumstances. When established members of the community such as representatives of government at international conferences [and] senior scientists from national institutions indicate the gravity and urgency of the global situation, including deforestation, is it equitable for the courts to impose injunctions that were traditionally an equitable remedy to prevent irreparable harm on those who try to prevent irreparable harm?

The courts have not yet made a ruling on their application.

NO
NO
NO

5

CONVICTION AND DISBELIEF: MASS TRIALS—2

THE MASS TRIALS CONTINued before different judges, and the cause produced a quality of oral expression seldom seen in this century. Defendants of all ages and from all walks of life stood before the courts with dignity, describing the myriad reasons for their actions at the bridge in Clayoquot Sound. Whether they spoke of forestry or

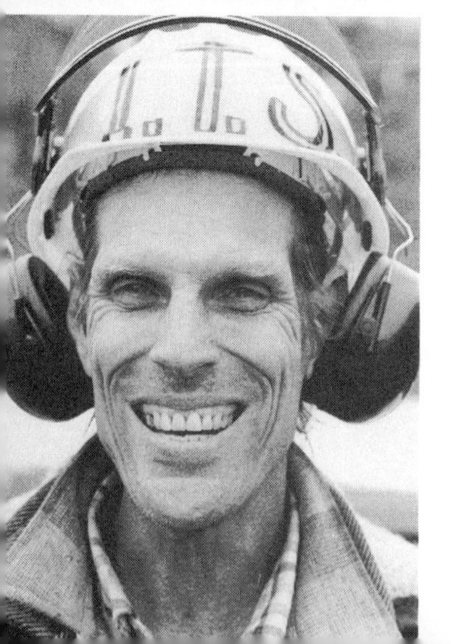

democracy, motherhood or spirituality, the urgency of the global environmental crisis or the philosophy of nonviolence and civil disobedience—all were supremely eloquent. We begin with the legendary eco-forester speaking in his own defence.

THE TRIAL

MERV WILKINSON: Milord, I am in this courtroom today charged with criminal contempt of a court order. I do not consider that any of my actions on the day of August 9, 1993 were criminal in any regard. Therefore I am pleading not guilty.

I did not on August 9 do any damage to the property of anyone; cause any bodily harm to any other persons there, including the police; and did not in any way use language offensive to anyone.

By objecting in a peaceful manner to the activities of MB, a multinational corporation with more convictions for breaking forestry laws and contravening more regulations in their operations than all the defendants in this case put together, I do not consider that I am a criminal.

The road that I was standing on has been paid for by the people of British Columbia either by direct grant or subsidized by ridiculously low stumpage rates. Therefore I was in truth standing on my own property because I as a citizen have helped to subsidize MB logging roads for several decades.

If there is indeed any measure of contempt, it is not for the courts as such, but for the company that devalues my country, breaks its laws at [the company's] convenience, manipulates my elected representatives and then uses professional truth-benders (PR experts) to brainwash the population (for example with the "Forests Forever" campaign).

What premier in his right mind would step in and agree to shoulder at public expense the costs of enforcing a very questionable corporate court ruling? There has to be manipulation present.

I believe that the record of activities over my lifetime in and on behalf of my fellow Canadians should prove to you, Milord, or any other worthwhile person, that I am indeed a solid citizen. I stood along with scores of

like persons on that logging road.

I grew up with people and principles that told me clearly that if I saw a crime being perpetrated I should do everything in my power to stop it. Failing that I should further work to bring the culprit to justice. I still believe this.

Now at the age of 80 I am being told that when I see a vandal destroying my property, a thief taking my bank account, and at the same time stopping to rape the environment, that I am supposed to help by not getting in the way. Why? Because he has a slip of paper allowing him to do virtually anything. Milord, I simply do not buy into this concept of justice.

As a witness and a spectator during this set of trials I have heard people being told they should have gone another road, should have taken other action. Milord, I will hand you this group of documents typical of literally dozens and in some cases hundreds of representations to the proper people and to the press, commissions, sittings etc., on the very values that resulted in the Clayoquot blockade, all of which have been ignored completely or replied to in a manner that was little credit to the government, industry or any other of the forestry groups.

There is a point—there always comes one—at which action becomes the only possible alternative.

We are fortunate indeed that the high calibre of the people on the Clayoquot blockade has so far precluded the use of violence. These people, faced with ridiculous slander, have remained nonviolent. This is more remarkable because on almost every other continent people faced with as grave an issue have chosen violence. Let's not push it.

It has been well said that the forest industry of B.C. was born in corruption, raised in corruption, and seeks to maintain that corruption. We have good reason to know this.

First we took the land of the First Nations and the timber with it.

In the early 1900s, T.D. Patullo gave the Powell River Company half of Graham Patullo Island tax-free for 90 years. The Haidas were never consulted.

Robert Sommers later got 18 months for gambling away a timber licence.

URNABY ARTS COUNCIL

I am the operator of a forestry [business] that has harvested timber for 45 years off the same land and still has the forest.... Now, at 80, I simply must defend what is left of my country from the multi-nationals of vandalism.

William Bennett allowed the companies to police themselves.

Is it any wonder that you have some 800 people arrested for trying to stop this type of forestry?

Milord, it is not necessary to destroy the forest to extract timber. It is a matter of method. Many systems are available that can operate in a manner that is acceptable to [the] environment and environmental people.

MB, like all the other companies, knows how I was present in a group of people some time back where a top forester for MB told us, "The reason we will not do alternate forestry here is that we would be admitting that we know how." This statement is well-known to many.

The refusal to do a proper job in the interests of foreign investors is unacceptable.

Proper forestry is more a matter of using more labor and less equipment. Skulduggery by the unions and operators has blinded the average worker and public to this fact.

There are probably as many reasons for people standing on the logging road as there were people who did so. Very bona fide to all of them.

I stood on the road because my conscience as a Canadian who loves the country he was born in compelled me to do so.

Every 322 hectares of clearcut forest in B.C. puts one forest worker out of work for at least 150 years. Foreign foresters basing their assessment on their own experience refer to our methods as a complete disaster for the forest and the environment. I cannot but concur in their assessments because I too am a forester....

I am a grandfather and great-grandfather and as such I would be traitor to my family not to have challenged an injunction that guarantees the right to destroy.... I am the operator of a forestry [business] that has harvested timber for 45 years off the same land and still has the forest....

Came the war in 1939...not being a pacifist, although having low regard for militarism,...I joined the PCMR, 31st Company and became Signalman 2nd class....

At present I am Field Instructor in Forestry for two universities, three colleges in Canada and one in the U.S.A. Consulted and called upon by the Audubon Society. Field guide for dozens of school classes. Assistant and

consultant for Community Forest projects in B.C., Ontario and three U.S. areas. Author, writer and consultant producer for videos on better forestry practices.

My property is now under study by German foresters and their staff members, as a pattern for rebuilding their forests destroyed by CLEAR-CUT AND PLANT.

Now, at 80, I simply must defend what is left of my country from the multinationals of vandalism. My grandchildren and my great-grandchildren must not be destroyed by corporate greed and bad laws.

TRIAL 19, FEBRUARY 18, 1994

SUE WHEELER: For the past ten years, off and on, I have been studying and attending workshops and conferences to learn more about forest ecosystems and forestry practices. I have collected a shelf of books...and have extensive files on sustainable methods of forestry—alternatives to clearcutting....

[MacMillan Bloedel] had little motivation to listen to other users' needs, or to work toward compromise. This practice is called "talk and log" and it...was the cause of the breakdown of the peaceful, participatory Clayoquot decision-making process....

So I was particularly upset at the Clayoquot decision last April, which set aside a few areas that would not have been logged in any case, and allowed clearcutting to continue as usual. At that time and through the summer I wrote letters and made phone calls to members of the government and signed and circulated petitions to express my displeasure about the decision. My own MLA, Leonard Krog, said in a widely published letter that environmentalists who objected to the Clayoquot decision were pissing on the tent. In an article in a Parksville newspaper he said he had no time for people who objected to the Clayoquot decision, essentially telling me to shut up and go away.

What is a responsible citizen in a democracy supposed to do with that...?

I went to the Clayoquot Sound area for a couple of days in late July. I went to the beach and then to the protest site as an observer on the side of the road.

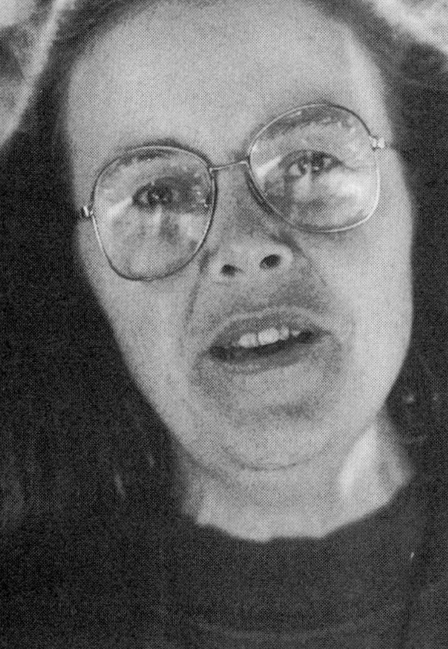

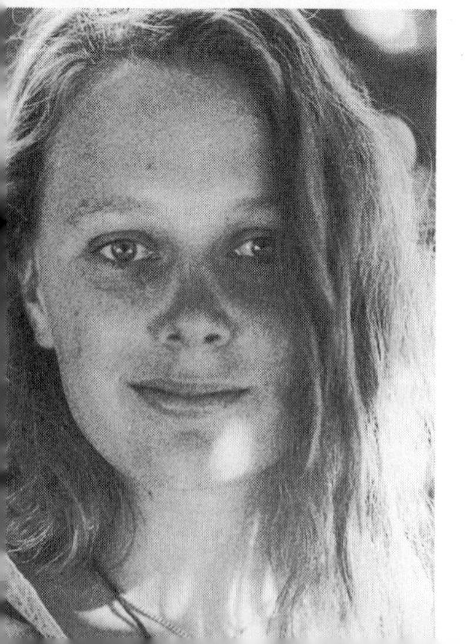

I went back to the Clayoquot Sound area on September 13....

That evening I thought about how valley after valley, river after river, habitat after habitat has been destroyed. I have three children and three grandchildren who all live on Vancouver Island or on an island off the east coast of Vancouver Island. This is the only forest they will have.

I thought about the fact that 93 percent of the original forest on Vancouver Island is gone, and here was this company, with its abysmal environmental record, being handed the right to clearcut part of the remaining seven percent and I thought—enough! So I stood on the road, as the only alternative left to me. I never imagined that I could prevent MacMillan Bloedel from putting in a full day's logging, and in fact I did not. It was a two-minute personal act of conscience.

◆

PETER HOLMES: Hundreds of conscientious, thoughtful, good people of all ages and strata of society have been arrested for peaceful expression of their ideas....

From these trials has come a public perception that something is wrong. Very many people in the general public—probably most people—recognize that the arrestees are not ordinary criminals and indeed are not criminals in any ordinary use of the word.

I believe that these trials themselves are doing at least as much to bring the administration of justice and the courts themselves into disrepute as, and probably quite a bit more than, the original actions of the arrestees that triggered these trials.

There is no doubt that the courts can slow down and possibly stop these breaches by increasing the likelihood and severity of sentences. Using the authority, power and might of the law will have this desired effect, but at what cost?

One effect, as I have suggested, will be that ordinary citizens will have reduced respect for the courts, and the law, because they perceive that the courts are being too severe and are responding harshly to basically good, valued citizens, acting peacefully and responsibly in the public interest.

Voting is not democracy! Much more is required of citizens. Non-voters have (excessive) influence between elections as well as great influence on

each election. A small minority close to the centre of power enjoy enormous privilege.

The emphasis on preserving the rule of law is misplaced and overemphasized. There was no real threat of violence, breakdown of law and order, insurrection, rebellion etc....

The law is not fragile. It is kept strong by active, vigilant citizens. Respect for the law is not gained by power, or threatening punishment. It is gained and kept by being seen and seeming to be fair and kind. People and animals do not act well from fear of punishment. Love and acceptance move them more, and make them stronger and happier.

◆

CAROL JOHNSON: Your Honor, I think we all sense that we are caught in a tragedy here, compounded of all the usual ingredients except perhaps heroism. We are all feeling, I suspect, ill at ease and miscast in our assigned roles. We are all waiting for the voice of wisdom; we would settle for the voice of common sense, but what we are hearing on too persistent a note to be ignored is the voice of paranoia. (We must put down these rebellions: we must deter these unseemly displays of opinion *because they unravel the fabric of democracy!* or because they somehow besmirch the rule of law!) Paranoia can occur in the best of families and in democracies that have failed to arrive at maturity. It deserves our pity and concern, it invites therapeutic intervention, but it does not evoke the unqualified obedience of reasonable persons.

Civil disobedience is evidently not well-understood in these parts. The rule of law is not flouted but affirmed by the symbolic gesture of concerned citizens peacefully expressed. Democratic society is not threatened but revitalized by such considered expressions.

Those of us who stood on a public bridge accessing public lands did so as a profound expression of the seriousness with which we take the obligations of citizenship. Neither the public nor the aboriginal inhabitants of these lands were ever consulted [about] their disposition. The injunctions routinely supplied to the corporation in question—whose practices have never been adequately supervised by any government though they continue to be an international scandal—these injunctions give every appear-

ance of improper manipulation of the law against the public interest, specifically to intimidate. Thus the peaceful presence of protesters represents a timely and a necessary intervention. We responded to a moral imperative.

Our intention was to avert a clear and indisputable danger, in this case to the very biosphere, for the programmatic destruction of the last old-growth forests at Clayoquot Sound seriously compromises our planet's capacity to sustain life.

The court does not wish to hear the argument of necessity. It would prefer to think that this exercise had nothing to do with trees.

History will certainly vindicate the Friends of Clayoquot Sound. History will understand that we were impelled by reason rather than criminal impulse.

Meanwhile there is ample evidence to demonstrate the moral obligation of citizens to respond to situations of this kind when the law does violence to the common good. We must recognize that the laws are only as good as those who write and administer them, and that since laws are composed by men who are capable of any venality or aberrant distortions of judgment, this sort of thing can happen.

The infamous Nuremberg Laws of 1935 were promulgated by an elected government of thugs and psychopaths. Apartheid was the law of elected governments for quite some time in South Africa, despite its repugnance to reason and the international community.

Huck Finn had a struggle with his conscience in deciding to help Jim escape from slavery. The laws at the time in states along the sides of the Mississippi made it a criminal offense to help a slave escape from his owner. Huck was worried by the fact that Jim said if he couldn't save up enough money to buy his children back, he would steal them from their current owner. Huck decided to go ahead and break the law and Huck was right in doing so.

No law, however well-meaning, if ill-judged, requires those citizens in possession of common sense in the face of potential global catastrophe to abnegate all reason in deference to a government in unseemly collusion with a powerful corporation, with or without the protection of injunctions

I believe that these trials themselves are doing at least as much to bring the administration of justice and the courts themselves into disrepute as the original actions of the arrestees that triggered these trials.

whose intent is to intimidate against the public interest.

The dangers to which we refer have been amply addressed by disinterested scientists who are not in the employ of any corporation for Public Relations. Their warnings have been made public for at least twelve years. *The National Geographic*, scarcely an inflammatory periodical, described the horrific results of clearcutting in the area in question three years ago.

So long as public officials in violation of the public trust respond to crises created by their own bad judgment with damage control and intimidation, there is a need for conscientious citizens to intervene as many are doing at Clayoquot Sound, peacefully. But with an urgent agenda, to save what remains of a fragile ecosystem.

Unlike Huck Finn, I know that what I did was right and necessary. And now, having fulfilled the requirements of my moral imagination, I commend myself, Your Honor, to yours.

Editorial note: The media reported that Carol Johnson, the 64-year-old professor, fitted with her grey ankle bracelet, will be wearing her metaphorical ball and chain for 28 days. Most of the protestors in her sentencing group got 14 days. Justice Richard Low told Johnson she was intellectually arrogant, gratuitously insulting and in dire need of deterrence.

Johnson shrugs off the punishment as something she can handle. But she bristles at what she saw in court: "I find it really repugnant to see young kids of college age making abject apologies for criminal behaviour to the court."

◆

BRYCE LYON: ...People who used or advocated the use of civil disobedience with or without a clear commitment to nonviolence were, for example: Jesus Christ, St. Thomas Aquinas and the Quakers, slavery abolitionists, suffragette Susan B. Anthony who was arrested for voting in a presidential election in 1872.

H.D. Thoreau argued that he did not owe allegiance to a government that captured runaway slaves and waged war on Mexico to expand its area of slavery. He was jailed.

Gandhi led Indian people using such nonviolent acts as strikes and protest marches to free themselves of British Rule....

Martin Luther King and other civil rights workers in the '50s and '60s deliberately violated Southern segregation laws as a means of fighting racial injustice. In the North, black people used civil disobedience to press their demands for open housing laws and equal employment opportunities.

[The] Vietnam war (1957-1975) had many opponents.... The civil disobedience campaign ended this immoral war.

In Canada, we have our own history of civil disobedience: Mennonites, Hutterites, Doukhobors, Quakers, etc.—religious communities that came to Canada upon the express undertaking that they not have to bear arms, yet were forced to defy the law.... The protests during the Great Depression of the '30s, particularly the March on Ottawa.

In the U.S. and Canada, it has been necessary to use civil disobedience to...gain respect and recognition of human rights for non-property-owning white men, black men, women, native people....

People around the world are using civil disobedience to end immoral wars and preparation for warfare. They are working to end the nuclear madness (H-power and bombs) that threatens to exterminate life and render our planet uninhabitable for tens of thousands of years.

Civil disobedience is also being used to try to stop environmental and ecological destruction of such scale and extent that it threatens life on Earth, at least as we know it.

Frederick Douglas, slave abolitionist, said, "If there is no struggle, there is no progress. Those who profess to favor freedom, and yet depreciate agitation, are people who want rain without thunder and lightning, they want the ocean without the roar of its many waters. Power concedes nothing without a demand. It never did and never will."

BEFORE JUSTICE COWAN, FEBRUARY 14, 1994

JANE TAL: I would like to start with a quote by Mike Harcourt on June 13, 1989: "The problem in B.C. is that this Socred government has a civil war, valley by valley, because they set it up so loggers are fighting environmentalists. Parks and wilderness are being encroached upon. The Socred game is to have those people at each others' throats. We have got to stop all that...."

We arrived at the Peace Camp after visiting the West Coast of Vancouver

Island for the first time.

We walked through the Rainforest Trail and realized that the rich, fertile soil was only inches deep. Once stripped away it leaves a powdery silt.

We then went to the lookout at Radar Hill. We turned the telescopes away from the sea and inland towards the mountains. I did not need to be an expert to realize that what had occurred there was, for the short *and* long term, devastating. A form of passive suicide. I felt that what I had learned constituted an emergency....

I come from New Brunswick where we have had extensive experience with tree farms for over 200 years, since cutting our own large trees....

I would like to end with a quote by H.R. McMillan at the Royal Commission on Forest Resources in 1956. He was warning against the Tree Farm Licences:

> "It will be a sorry day for British Columbians when the forest industry here consists chiefly of a few very big companies holding most of the good timber and good growing sites to the disadvantage of the most hard-working, virile, versatile and ingenious element of our population, the independent logger and the small millman. A few companies would acquire control of resources and form a monopoly. It will be managed by professional bureaucrats. Fixers with a penthouse viewpoint who, never having had rain in their lunch buckets, would abuse the forest. Public interest would be victimized because the citizen business needed to provide the efficiency of competition would be denied logs and thereby be prevented from penetration of the market."

◆

ANGELA HEYWARD: ...I believe I am directly responsible to provide a sustainable future for [my children].... My actions on September 14 were a direct result of my personal beliefs and convictions [about] my heritage, my future and my children's future....

MacMillan Bloedel, [which has] been convicted 25 times between 1969 and 1991 of criminal charges directly related to their logging practices, has continued to work in non-compliance with the law in over 60 cases. Their practices are currently being investigated for criminal charges within Clayoquot Sound. Your Lordship, for these reasons I believe I was not obstructing justice but rather the injustice caused by MacMillan Bloedel's

"If there is no struggle, there is no progress. Those who profess to favor freedom, and yet depreciate agitation, are people who want rain without thunder and lightning..."

continued [flouting] of the law....

The government spent millions of our taxpayers' dollars in attempts to downplay these protests instead of dealing with them. Meanwhile in every town in British Columbia people were actively taking sides, to the point of physical and verbal conflict. MacMillan Bloedel, instead of complying with their obligations, enforced the injunction and attempted to downplay their non-compliance, using loss of employment threats and organizing active anti-protest groups.

My only choice was to bring these travesties to the direct attention of the courts....

◆

PATRICK O'ROURKE: I would like to preface my statement by saying I come from a family with a strong commitment to social justice, human rights and environments.

During the past two decades we have been involved in many peaceful demonstrations against nuclear weapons in North America and Europe and have walked many thousands of miles on international peace walks across the United States and Europe, and our children have spent many seasons tree-planting on the clearcuts from the Charlottes to Ontario.

I do feel, as everyone does, the things we value in society have to be protected and feel that the natural impulse to safeguard them may be inherited as it may well be with all of us as we proceed on this journey through life. In my own case I come from the Gaelic family O'Rourke...of Ireland.... Brian of the Ramparts O'Rourke, during the winter of the Spanish Armada, took it upon himself to shelter some of the Spanish sailors who had been shipwrecked off the Irish coast. Due to the fact that they would immediately be hanged, he refused to surrender them to the Crown (at the that time worn by Elizabeth I).... Brian was...eventually beheaded.... While I have a head on my shoulders, I have to make my feeling known in as nonviolent a way as is possible, as I did on the morning of August 9, 1993....

Clearcutting is an ecologically disastrous practice which destroys the integrity of the forests ecosystem and leads to rapid erosion of the soil. It takes 100 years at average soil building rates to develop one eighth of an inch of soil.... I believe that the books *Wildwood: A Forest for the Future* and

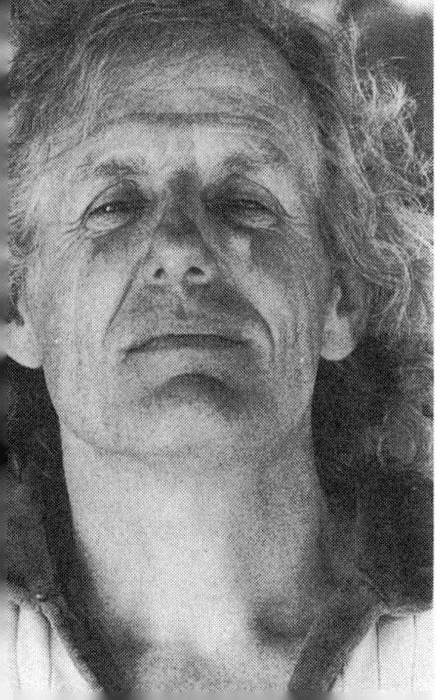

Keeping the Forest Alive regarding selective and sustainable logging on the 136-acre forest of Merv Wilkinson should be mandatory reading for all foresters.

Multinational forest companies' cutting rights have already turned 20 percent of Clayoquot Sound into an industrial plantation and now 70 percent of the remaining forest of commercial value is open to clearcut logging. Trees are the oldest and tallest living things on this planet. Without forests the Earth cannot breathe (her lungs are trees) and old-growth forests are the Earth's most important storage system for carbon.

During one spring and summer in the late '80s, [I] and my four children were all out tree-planting. We all generally agreed that you could walk...from one side of the valley to the other without touching the ground, walking on perfectly good timber left to waste....

An operator can now sit in the comfort of his cab and manoeuvre two massive claws around a centuries-old tree and log it in a few seconds. In a single day a faller/bucker can cut enough timber to fill 27 logging trucks.

The province of B.C. employs fewer people per cubic metre of forest cut than any other nation in the world....

In the past three years B.C. has cut enough timber to load a convoy of logging trucks 75 miles long every single day, and the Ministry of Forests calculated the harvest to be 25 percent above the sustained yield. Seventy-five percent of trees cut are mature or ancient forest. The United States limits clearcuts to 16 hectares but here in B.C. they go anywhere from 30 to 300 hectares....

I sometimes have had to wonder if the prosecution of everyone at no expense to MacMillan Bloedel could be a lot more than self-serving for them because then you know that if they continue to carry on logging in Clayoquot Sound, it will extend the protest, causing more discontent and confrontation amongst the electorate in British Columbia and turning more public opinion against the existing B.C. government. [The result would be] that a new administration...even more favorable to the forestry industry would be returned at the next B.C. election, giving the forestry companies a free range to clearcut the province from north to south, as we had for two decades under the Social Credit government.

◆

KAREN LANG: There is a story that is told about Buddha. This story had a tremendous impact upon me a few years ago and contributes to my reason for standing on the road at the Kennedy Bridge.

There was a man who was a mad murderer and had taken a vow that he would kill 1,000 people because society had not treated him well. From every person he would take one finger and make this into a rosary around his neck. His name was Angulimala, The Man With A Rosary of Fingers.

Angulimala met Buddha. Angulimala had a stir of compassion for Buddha but there was this vow to fulfil. Buddha said, before you kill me, do one thing, just the wish of a dying man: cut off a branch of this tree. Angulimala swung his mighty and high-tech sword against the tree and a branch fell down.

Just one more thing said Buddha: join it, again, to the tree.

Angulimala said, "Now I know, perfectly, that you are mad. I can cut but I cannot join."

Buddha began to laugh and he said, "When you destroy and cannot create, you should not destroy, because destruction can be done by children. There is no bravery in it. This branch can be cut by a child, but to join it a Master is needed...."

Well, we are not talking about human hands here. We are talking about beautiful, magnificent, ancient trees and a whole ecosystem that is based around them....

◆

JILL SMITH: We all have a responsibility to the Earth—when we hurt the Earth, we hurt ourselves. When we hurt each other, we hurt the Earth. When we hurt the Earth, we hurt each other. My heart cries at the hurt we are doing to the Earth, our mother, our soul....

I want those who have not felt the Earth, who have not listened to her cry, who have not seen the color of her hair and clothes as they change with the season...that have not smelt the breath of the earth after the rain...to share them with me, and to know too that we must...hurt the Earth no further that she may live, else she may die.

◆

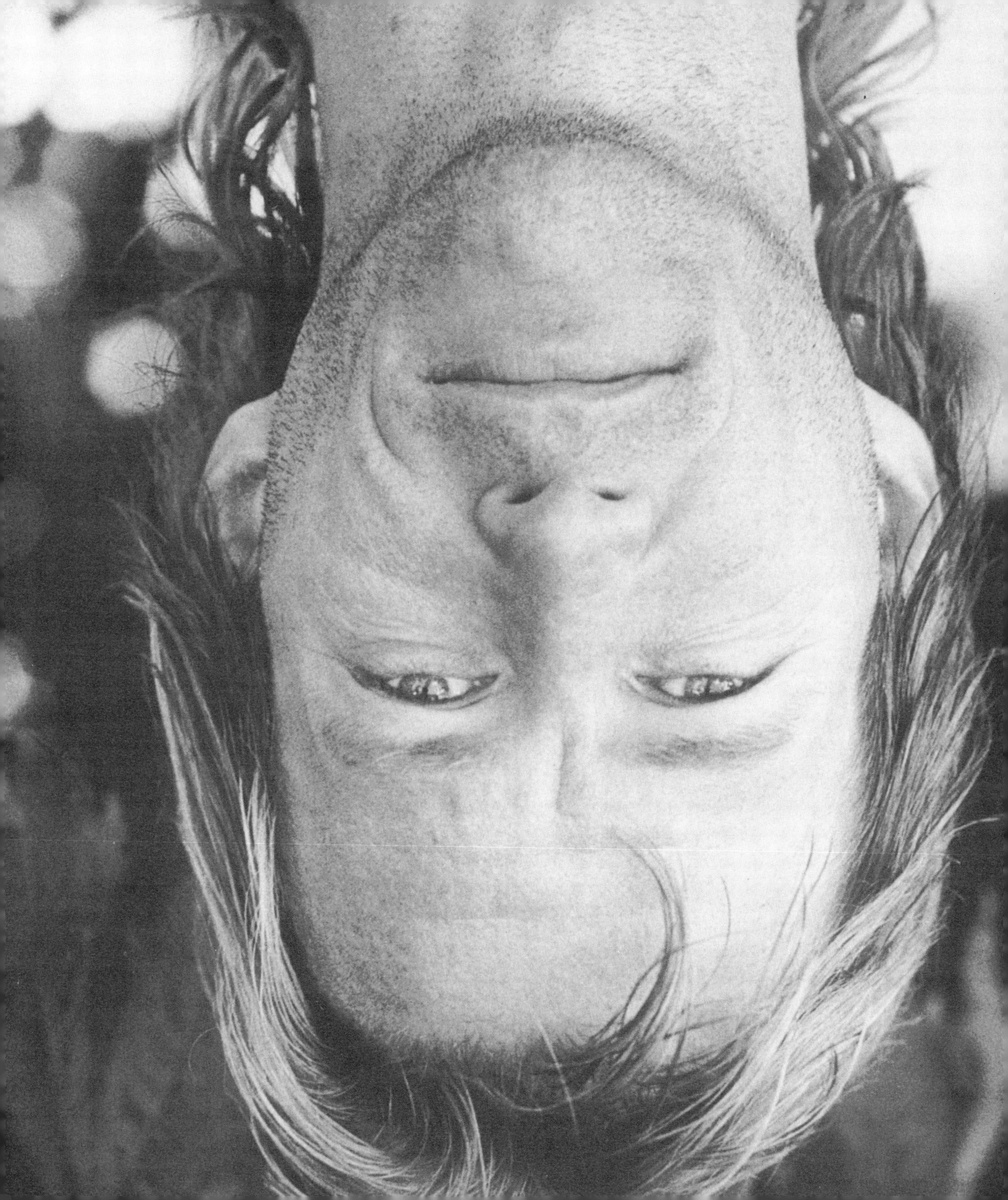

SUSAN COGAN: My Lord, you said that you do not come into this courtroom in a vacuum, and that you have an obligation to rule as your brother judges have done before you. Like most Canadians, I have no familiarity with court procedures. I assumed that each judge would be obliged to view things afresh. When I stood on the road, I was putting my faith in the courts.

I thought that the court would see from the great number of people standing on the road that things needed to be looked at again, that the people could come before the court as an emergency measure. That is why I am here My Lord, not to disparage the court.

Human activity is damaging the Earth gravely.... I believe Your Honor said it to be a matter of opinion....

So I put it to you, what if it is more than opinion? What if we are right? What if cutting down the last remaining rainforests of Vancouver Island will directly contribute to our demise? We may not see the results for years, but we cannot paste the forest back up once the trees are felled. I greatly fear the cutting down of so much pristine rainforest. It seems there is a race in the forest industry to cut as much as possible before the number of people who object grows too big....

The clearcutting of Clayoquot Sound is akin to the burning of the Great Library of Alexandria before anyone read the books. So much is unknown of rainforests.

◆

MARK DAVIES: ...The court system operates through a process of intellect, but in my experience erudition is not necessarily where truth and wisdom lie.

I believe the heart and compassion to be the true seat of knowledge and that truth is very much a feeling experience. For example, if people had not acted as they did in the suffragette movement, claiming women's democratic right to vote, or in the demonstrations against the Vietnam war, or in Martin Luther King's actions to secure human rights and voting rights for blacks, sanctified oppression would still be a part of our society today....

It is a dark time in the history of Canada when we, the people of this

What if we are right? What if cutting down the last remaining rainforests of Vancouver Island will directly contribute to our demise?

beautiful country, are jailed for our efforts to protect, respect and show our love for it....

One might adapt a comment from Oscar Wilde that MacMillan Bloedel knows the price of everything and the value of nothing....

When anything is objectified and only seen through the eyes of profit, it naturally ignores and, in the case of the forests, destroys its intrinsic value. The varied forms of species and elements were not put here solely for our manipulation and use, as certain corporate forces would have us believe. It is becoming clear that we need to develop a new relationship of grace, dignity and respect for all things, inanimate and animate, if the human race—at this point racing to its demise—is to survive....

To walk amongst trees in an old-growth forest is to walk through an awesome beauty that is unmatched by anything man-made, and that once destroyed can never be replaced. How can we ever justify borrowing from our children something that we can never repay...?

My Lord, I would like to refer to a comment that you made on Tuesday, January 11, regarding stoplights and anarchy. I believe you were describing the potential problem that would occur if everybody disobeyed the traffic signal and proceeded at their own whim. I would like to borrow your metaphor and apply it to MacMillan Bloedel.

Considering that they have 23 convictions to date and I do not know how many more pending, the members of the community are justly confused as to why MacMillan Bloedel still have their driving permit....

◆

TARA MACLEAN: ...I am speaking now to try to help you to understand that my actions were not of a criminal nature....

On September 1st, the day before I got arrested, a busload of Victoria business people on their way to Clayoquot Sound were blockaded by a group of angry loggers and by the Share group. The public highway was blocked for hours and not one arrest was made. These business people were held captive there for hours. They were cursed at and harassed and not one single arrest was made. The Share group rocked the bus and let air out of the tires and not one arrest was made. The police just stood there.... I went that day to Clayoquot Sound to see what was going on. I had been

well aware of the logging practices of the company...but I had no idea of the conspiracy between the government and this company. The police said that they were afraid to start anything at the Share blockade on September 1st. So if you are volatile and angry and violent, then the police will not arrest you, but if you practice nonviolent, non-cooperative passive resistance, then they will arrest and jail you....

John Kennedy said that those who make peaceful revolution impossible, make violent revolution inevitable.... If it were not for peaceful protesting and civil disobedience, then many of our rights and freedoms would never have been achieved. For example, at one time you had to be a white male who owned land to vote, otherwise you did not count. If I had to cross a "no trespassing" line to help someone in trouble, I would not even think twice....

◆

SHELDON LIPSEY: ...I am a translator by profession. I worked for the Secretary of State Department in the federal government for five-and-a-half years, as a self-employed professional for the last thirteen years....

As a self-employed individual I have suffered as much as anyone from the recession. When I was arrested this summer, I considered the expense of returning to British Columbia for trial and decided it was part of the price I would pay to make a statement of my political and personal beliefs....

As I returned to Vancouver Island frequently over the years,...I realized clearcuts that used to be hidden were now visible from the road. Cathedral Grove..had become a tiny oasis in a desert of spindly second-growth. At the same time...jobs were disappearing steadily.... Why wasn't forestry sustaining the economy...? Wasn't MacMillan Bloedel's motto "forests forever"...?

The facts, horrifying as they are, pale in impact compared to the actual experience of seeing a clearcut close up. One day last summer, I had the opportunity to visit the active logging area in Cold Creek. From a safe distance, I and others in the group looked down at the devastation in the valley [at] clearcut slopes right down to the creek, or rather where the creek used to be. Debris lay everywhere—felled trees left as waste, old cables,

broken equipment. There were no words to describe the horror and shock of seeing such destruction. We stood a long time, some of us cursing silently, others weeping.

Standing there staring at the dead landscape, I had a moment of utter terror. There I was, catapulted ten or fifteen years into the future, and this was what our world was going to look like. All of the wilderness turned into desert, rivers too laden with silt and debris to sustain life. No wild and natural places left to escape to.... I realized then just how precious and vulnerable Clayoquot Sound was. Too precious and too vulnerable to be shared or chopped up into little bits that could not survive.

When I learned of the B.C. government's decision in April 1993 to allow logging in Clayoquot Sound, there was no doubt in my mind that the democratic channels we're all supposed to trust had failed.

...Civil disobedience is not simply about the right to vote. It is a refusal to comply with laws that are perceived as unjust—in the case at hand, laws that protect a private company's financial interests in cutting down trees but fail to protect the public's interest in preserving the country's natural heritage, as well as land to which the First Nations have never given up title.

As Canadians, I and others were protesting an unjust situation that arose despite the fact the people of B.C. had the right to vote—indeed, precisely because they voted to elect a new government on its promise to better protect the public interest and the environment, and watched as the government betrayed their trust.

Mr. Justice Bouck in his Reasons for Judgment offers this unusual description of democracy:

> "Sometimes even the best arguments fail to persuade. At other times, democratic governments pursue long-term policies that do much harm.
>
> "Years after they are found to be misguided. But these faults are the price we must pay for democracy because the alternatives are worse."

I firmly believe there is no alternative worse than wilful destruction of the planet. If Mr. Justice Bouck were asking me to accept such a concept of an imperfect democracy, my answer would be simply—never.

...There is an even more important aim of dissent: to convince those in authority of the validity of the dissenting opinion.

By standing in this court, I hope to gain not only an acknowledgment of the political nature of the conflict, but a declaration that the public interest is not being served by criminalizing eight hundred people, that the issue should be handed back to the legislative branch with the admonition that it will have to do better, that criminal convictions are no way to resolve a private dispute between the government and a special interest (MacMillan Bloedel) on one side and dissenting citizens on the other.

Such a declaration from the court would realize Gandhi's concept of *satyragraha*, the truth force that transforms relationships between citizens and the state—not only resolving the conflict but restructuring the situation that led to the conflict, and transforming people and attitudes in the process.

With the greatest respect for the court, I remain confident that the court will make that declaration in pronouncing sentence.

◆

In engaging in this act of civil disobedience, the statement I make is that I believe so strongly that the forests of Clayoquot Sound should be protected that I am willing to suffer personal consequences to demonstrate my belief.

VIVIAN CHENARD: ...Education and personal example are the essential ingredients to effect social change and I realize that it is important to work through established and legal channels. However, I am alarmed at the rate at which humans are destroying their home place. When my grandchildren ask what I did to halt this destruction, I want to say that I did more than write a letter....

In engaging in this act of civil disobedience, the statement I make is that I believe so strongly that the forests of Clayoquot Sound should be protected that I am willing to suffer personal consequences to demonstrate my belief....

◆

REVEREND BROWNLEE OF SASKATCHEWAN: "In the beginning when God created" is a familiar beginning of the Bible. An alternative translation of that scripture, and the translation in the latest English version of the Hebrew Scriptures, is "When God began to create...." My Lord, the difference is significant. It points to the fact that creation is ongoing, not just

something that only happened in the past, something that is concluded once and for all.

No matter what God we worship, or what ideology we endorse, this fragile Earth, our island home, is of vital concern to all of us. The equation is simple: no living Earth, no living human beings.

The Hebrew people understood that equation. From their time in Egypt through their time in the wilderness, they looked forward to their time in the Promised Land. On their way to that promised land, God gave them, and us, a code of life that allowed successful living in community. A key part of that code was the understanding of our relationship with God....

God says..."if you don't take care, your land shall be a desolation and your cities a waste...." My Lord, I am a priest of the Anglican Church of Canada. My church has been concerned about the use and abuse of God's creation for a long time....

> "The Synod of the Diocese of British Columbia deplores the wasteful methods of logging in vogue on Vancouver Island and the Coast generally, and the consequent destruction of the valuable forest resources upon which so much employment depends, together with irreparable damage to the watersheds and spawning grounds; and further expresses the hope that early steps will be taken to enforce selective logging in order that a large measure of waste may be eliminated and immature timber saved from destruction."

That motion was passed over 50 years ago, in 1939. It was a plea to save our environment, and increase employment.

In 1975, the General Synod of the Anglican Church of Canada passed a resolution "that no new major industrial development in the north of Canada should be initiated on unsurrendered land until native land claims are justly settled, or terms governing that development are negotiated satisfactorily with the native peoples concerned."

While Clayoquot Sound is not in the north of Canada, the principle still applies. Clayoquot Sound is unsurrendered Nuu-chah-nulth territory. The land is not owned by British Columbia. It has never been given to MacMillan Bloedel. My Lord, the very land which I attempted to protect is not ours!

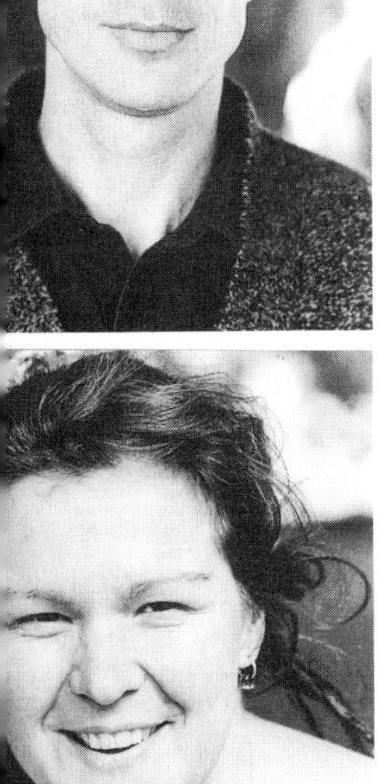

In 1989, the General Synod of the Anglican Church, in its frustration at

the lack of justice in the way we treat our native people and their environment, passed Act #62, which reads, "That aboriginal peoples and Anglicans and others involved in nonviolent direct action in defence of unsurrendered ancestral lands and their environment be supported."

The issues of justice surrounding us include a land claim, our responsibility for Creation, and blocking in defiance of a court injunction.

My Lord, I was ordained a priest ten years ago. Before that...I worked...for several...multinational corporations...[whose] prime purpose is to provide the largest profit possible. If they do not do this, their...chief executive officers usually lose their jobs.... MacMillan Bloedel...is there to log...in the most...profit-generating way. That way is clearcutting.... My Lord, this land could produce more profit, more employment, if it was logged wisely, not raped by clearcutting....

My Lord,...is it possible for you to use your good offices to try to bring an end to this fighting? Can you make a judgment that would encourage a stop not to logging, but to present logging practices while true and meaningful discussions resolve the issues now and forever?

My Lord, for me the issue is deeply religious, deeply theological; it is a question of the relationship of the Creator to the Creation. It is a matter of our responsibility to our Creator....

◆

COMMUNITY OF CONCERNED MOTHERS: A PRE-SENTENCE STATEMENT: We are a group of mothers who were arrested on the mornings of August 11th and 13th at the Kennedy River Bridge. Our ages range from 24 to 60. Some of us are new mothers, some are mothers of grown-up children. Esther, the mother of two daughters, is a counsellor at the Victoria Male Survivors of Sexual Assault Society. Ute, an environmental activist who has worked within a number of environmental organizations, is the mother of two teenagers. Joanne, who is sixty years old and the mother of one son, has also been long active in the peace and environmental movements. Nancy is a single mother of a 4-year-old son. Michelle has a three-year-old daughter. Ann, a clerical worker and the mother of two grown children, has been actively concerned with environmental destruction for many years.

Clayoquot Sound is unsurrendered Nuu-chah-nulth territory. The land is not owned by British Columbia. It has never been given to MacMillan Bloedel.

We come from a variety of spiritual backgrounds. Many of us are Quakers. Betty, a pediatrician, has two grown children. She has worked in Ethiopia—now treeless and barren—and has seen the devastating consequences of unchecked logging. Her children are proud to see her stand on this issue. Miriam and Carol also stood at the bridge as an expression of their faith.

Although we have different themes in our personal lives, our common experience of nurturing and loving our children led us to take action to defend Clayoquot Sound, one of the last remaining temperate rainforests on Earth.

The scientific community is telling us that we must change the way we relate to the Earth, or die. In [November 1992] the Union of Concerned Scientists...issued a statement entitled World Scientists' Warning to Humanity:

> "Human beings and the natural world are on a collision course," state the scientists. "Human activities inflict harsh and often irreversible damage on the environment and on critical resources. If not checked, many of our current practices put at serious risk the future that we wish for human society and the plant and animal kingdoms, and may so alter the living world that it will be unable to sustain life in a manner that we know. Fundamental changes are urgent if we are to avoid the collision our present course will bring about...."

We are frightened and angry about the potential mass extinction of species on this planet, and the seeming inability of our government to take preventive action. In deciding to violate a court order issued to protect the logging rights of MacMillan Bloedel, we had looming in front of us the vision of a devastated Earth outlined by 1,575 scientists. What kind of a world will this be for our children?

We did not want to find ourselves reproached in later years by children who asked us why, when the Earth was threatened, we did nothing. Being arrested and accepting the consequences gives a strong message to our children that we will do whatever we can to protect their futures. What is the use of raising our children with care, if the planet has no trees to provide them with oxygen...?

If our planet is to survive, we must each take personal responsibility for its survival, and we must model this responsibility for our children. We

hope that our children, seeing us standing up for our beliefs, will learn from integrity and responsibility.

In the words of the World Scientists, a new ethic is required—a new attitude towards discharging our responsibility for caring for ourselves and for the Earth.... This ethic must motivate a great movement, convincing reluctant leaders and reluctant governments and reluctant peoples themselves to effect the needed changes.... We require the help of the world's peoples.

We do not regret our stand on the bridge. We fear only that what we have not done will haunt us. Where there is no vision, the people will perish.

Nancy Powell	Carol MacIsaac
Ute Frank	Joanne Manley
Ann Hughes	Michelle Mueller
Crystal Betty Kleiman	Ester Muirhead
Miriam Lee	

◆

ERIKA FOULKES: I am a thirty-six-year-old registered medical laboratory technologist in Vancouver. I hold a bachelor of science degree in biology from the University of Victoria, and am currently working on a bachelor of arts degree in philosophy at the University of British Columbia. I was born in Vancouver and have lived for many years on Vancouver Island. This island is my home. I have never broken a law of any kind. Nor did I expect that I ever would, and I have felt greatly ill at ease about doing so in this instance.

Mr. Harcourt had promised to end the fragmentary valley by valley forestry decision-making strategy which has been the tradition in this province. Yet his government excluded Clayoquot Sound from the jurisdiction of the Commission on Resources and Environment.

I have watched the Clayoquot Sound Sustainable Development Steering Committee plummet into failure because of a cynical strategy which permitted irreversible logging operations to go on in contested areas at the very time that the fate of such areas was to be negotiated. I am satisfied that there is clear evidence that the government's land use decision for Clayoquot Sound is simply not good enough to protect the integrity of that

unique area.

MacMillan Bloedel is notorious for its repeated violations of the Fisheries Act and the Waste Management Act. Clearly MacMillan Bloedel is recklessly unconcerned about the results of its actions, and equally clearly, the provincial government is pursuing an agenda designed to facilitate a private corporation's pursuit of its own interests.

I am convinced that if we allow MacMillan Bloedel and forestry companies like it to continue to operate in the way in which they have been, we will simply have no intact wilderness left. Livelihoods will be destroyed. And we will still be too ignorant to estimate the extent of what we will have lost.

I have spent years writing letters to the provincial government, generally to no avail. I have attended protest rallies and marches. With the failure of Clayoquot Sound...I became disillusioned about the prospect for success within prescribed channels of halting the annihilation of our forests....

It is a slippery slope fallacy to suggest that the judicial system as a whole will be eroded, brought down by my single act of moral protest against MacMillan Bloedel. I do not wish to see the judicial system eroded. Our system of laws and the courts which protect them should properly be strengthened by well-considered acts of protest against a status quo that not only allows but encourages the unethical and frankly rapacious activities of one of society's members—in this case, MacMillan Bloedel....

I am disturbed that the police have seen fit to pass [over] evidence in their possession on the private forestry company involved in this dispute. I see no other way to interpret this than that the neutrality of the police as public servants has been compromised....

Although my actions at Kennedy River Bridge were deemed serious enough to warrant a charge of criminal contempt, the consideration of my case in the courts has not been treated with equal seriousness....

◆

YVON RAOUL: ...Nothing I have heard since or read has made me change my belief that the injunction was wrong, and that I and the others were right to risk considerable personal inconvenience in order to defend something we consider more important than ourselves. I deeply believe in democracy and I realize that respect for the law is essential for it to work. Yet

Being arrested and accepting the consequences gives a strong message to our children... What is the use of raising our children with care, if the planet has no trees to provide them with oxygen...?

our system has flaws and one of them is that the law can apparently be applied severely in some cases, and less severely in others, especially when large amounts of money or influence are involved.

The logging companies can afford to pay lawyers to defend their cause and image-makers to influence public opinion, as well as media coverage and advertising....

At the time of the blockade, the provincial government still had not published new, stricter guidelines. In fact that they have now been announced may have something to do with our peaceful protest....

The sentences given to some of those tried earlier shocked me.... I perceive them as the kind of intimidation associated with the repression of democracy rather than its protection....

Merv Wilkinson has shown that selective logging can produce more jobs over a longer period that clearcutting can. I have nothing against loggers and I apologize for the inconvenience that our action caused some of them. I hope that such protests will prevent them from finding themselves in the same position as the Newfoundland fishermen.

I also regret that this whole affair has taken up so much time in the court and so much public money, but this is not because of the protests but the choice of those who made this technically a criminal activity.

I am a teacher and a father; I certainly do not wish to provide a bad role model for either my own children or my students. As someone committed to defending the environment, for their benefit as well as my own, I could not with a clear conscience refrain from expressing my concern in the most effective way available.

I have appreciated the support of my family and friends, and I have met some admirable people through this experience, on both sides of the debate. I hope that whatever sentence I receive will not prevent me from fulfilling my duties at school.... Changes in our forest practices, or any change as a matter of fact, happen if enough people are ready to commit their time and their energy to do something like a peaceful road blockade. I compare my own action to a vote. I cast my ballot on the road, other people have done it before me and I hope many more will do it after me....

◆

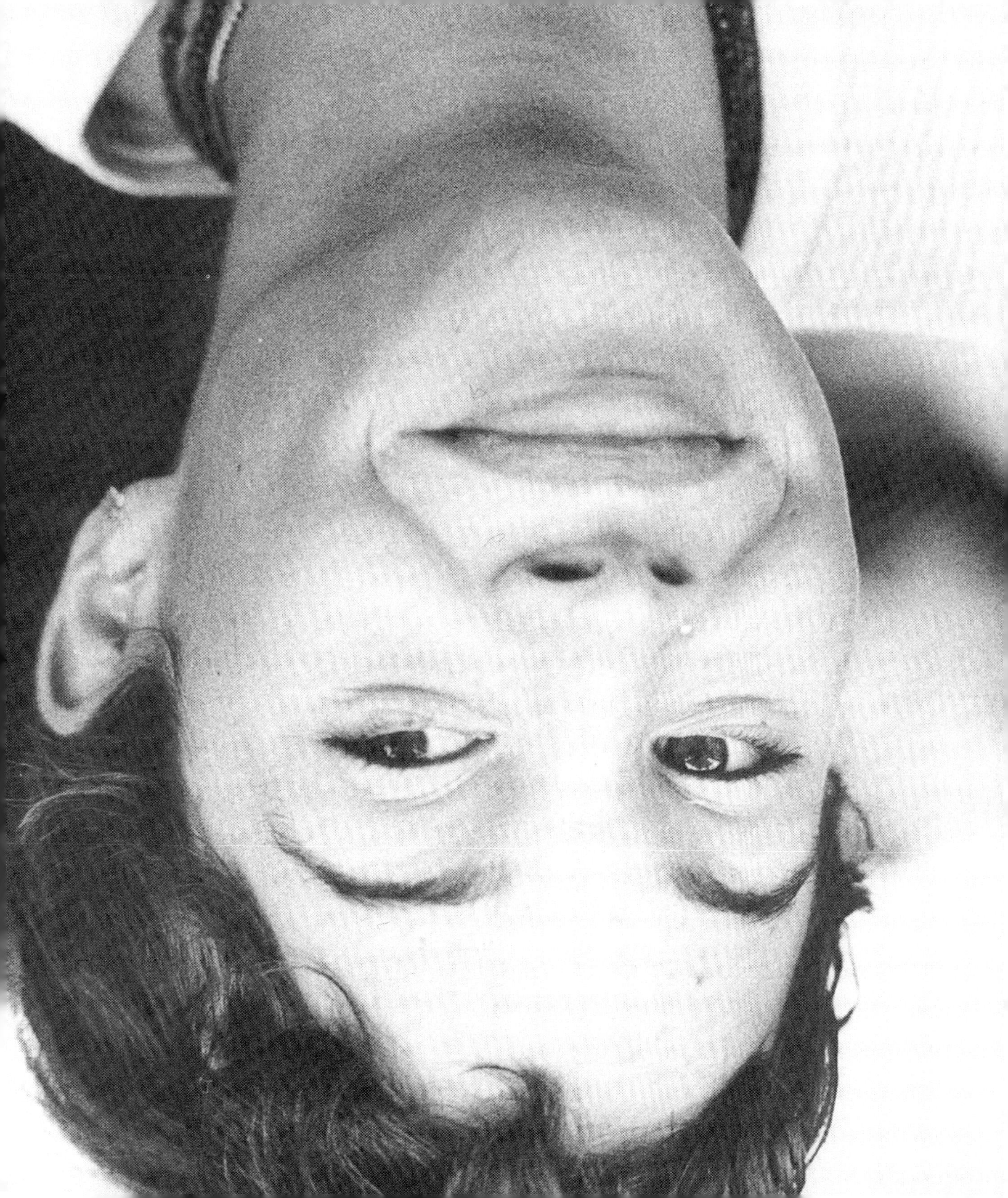

RINTJE RAAP: Please allow me to say a few things about myself—why I ended up in this courtroom, what I have learned from this experience.

For the past 22 years I have taught the fundamentals of chemistry to a few thousand college students and before that I spent seven years working in the pharmaceutical industry, trying to make better antibiotics. My chosen field of expertise, chemistry, has done great things for mankind but it also led to much environmental damage, especially in the last 50 years. We have all paid dearly for our chemical progress: for our plastics, our metals, our paper products, our medicines and our energy.

I see the forest industry in the total framework of our infatuation with technological progress and consumerism.... Sometimes lawful democratic actions do not work because the economic interests that control all of us, including our government, act against the long-term public interest.... You have declared me a criminal. I do not consider myself as such and neither do my family, my friends and my students....

What has this experience taught me? The most overwhelming realization it has given me is that we are all just pawns in a very serious game—we (the new batch of criminals), the RCMP officers, the defence lawyers, the prosecution, the government of our province and even you, My Lord.

The prime movers in the game are the big multinational companies such as MacMillan Bloedel. Their economic interests and power control us all....

◆

TIM TAYLOR: ...My Lord, I would like to thank you personally for lending humor, dignity and humanity to these proceedings.

I am a 47-year-old Canadian citizen with a wife and two teenage children. I immigrated to Victoria with my parents in 1957, left again in 1967 to study, work and travel, and returned in 1977. I have a BA in history from the university of Victoria and an MA in history from Cambridge....

I have camped and hiked this island since I was a boy and I am a horrified eyewitness to its devastation....

I am sorry that the Ministry of Forests is a captive agency of the forest industry, that even now it has made no complete inventory of lumber resources, that it relies almost exclusively on industry advisors, that it makes most of its decisions behind closed doors, and that it manages to run an

annual deficit despite huge industry profits....

I am sorry that MacMillan Bloedel is allowed to use its vast profits from carbon mining B.C. and from milking the federal and provincial governments through subsidies, grants, tax credits and tax deferrals, to establish a plywood plant in Alabama.

◆

ERNEST HEKKANEN: ... [Richard Bourne, process server for MacMillan Bloedel] says [MacMillan Bloedel agent Lorne] Dixon not only removed [RCMP Polaroid photos of protestors requested by MB and meant to be for MB eyes only] from the envelope, he passed them around to others in the restaurant.... For justice to seem to be done, charges must be brought against Mr. Dixon....

Dixon...said, "if they (meaning the RCMP) did not give us the Polaroids, they knew they would not get the video pictures...." If we allow the Wild West tactics employed by MacMillan Bloedel, the RCMP and the Crown to succeed, we are going to end up with our freedoms badly eroded....

◆

ERVIN NEWCOMBE: I have lived almost all of my life on this island. I fish, sail, canoe and hike here. For me the damage done by our current logging practices is not an intellectual exercise. I have grown up with it. It is so pervasive I used to think it was normal and that this was the price of decent employment levels. I accepted it as a necessary evil.

Over the last few years I have learned that this is not the case. We get much less employment, taxes and economic benefit per unit of timber than almost any developed country.... We have barely any secondary processing of wood products....

[Forest] companies have no commitment to their communities and indeed have in the past threatened mill closures or layoffs if they were not given concessions....

I see an understaffed civil service given impossible responsibilities and confusing if not conflicting direction.

I see the largest and most powerful forestry union in the province tacitly allowing massive layoffs due to mechanization and overcutting, whilst joining the forest companies in blaming the layoffs on radical, insatiable

conservationists....

I see MacMillan Bloedel convicted on numerous charges of damage to the environment. Usually these result in token fines representing a fraction of the profit gained by cutting a corner or ignoring a directive. These penalties serve as little deterrent. All too often these charges only arise after public exposure and pressure by environmental groups.

I sense that companies like MacMillan Bloedel are truly supra-national and operate in an anarchistic environment beyond the will, or at least the ability, of this government to exercise control for the public good....

◆

CHRIS MORRISON: ...It has been my impression that the firm of MacMillan Bloedel has a reputation for its greed, its massive destruction, multiple infractions, and its deceit. This is the firm that has hired the firm of Burson-Marsteller to be its ad agency.... Burson-Marsteller is the firm which was called in after the Exxon Valdez oil spill to smooth over the public outrage and this is the firm which MacMillan Bloedel has engaged. The very fact that Burson-Marsteller has been employed by MacMillan Bloedel leads one to assume that there is something they want to cover up....

So when I heard the news on CBC radio [about government purchase of MB shares just before granting them licence to cut Clayoquot], this was the last straw. I had to go and see for myself....

I had no previous knowledge of this area or the Friends of Clayoquot Sound. After I arrived at the Peace Camp I saw with my own eyes the terrible devastation. In the Peace Camp at the Black Hole, I saw that there was no life left there, neither birds nor animals. I went to Meares Island, saw my first huge old-growth tree and was filled with humility before it. For the first time in my life I knew the meanings of the words awe and wonder....

As I stood on the road in front of the truck...I was not thinking of the court, was not aware of the court, had no thought of the court.

I was aware, however, of MacMillan Bloedel. It was in fact an employee of MacBlo who ordered me to get off the road....

I am presently co-facilitating a 20-week group for Women Survivors of Sexual Abuse on behalf of an organization in Vernon, known as CASA—

I am most proud of three events in my life. They were all very difficult. Giving birth to my three sons, riding my bicycle from Vancouver to St. John's Newfoundland, and being arrested at Clayoquot.

Communities Against Sexual Abuse.

All of my clients and the people I work with are struggling with years of conditioning from a society which is based, in large part, on denial. These people are pushing past the facades which they have encountered in their lives and are learning to speak their truth.

My action at Clayoquot Sound was part of my own process of speaking my own truth and owning my own truth. I am content with the action I took and I am proud to be numbered amongst the brave people in this courtroom. It appears to me that our courts, although they uphold the law, are functioning in a void that perpetuates this social disease of denial.

Our planet is dying, our house is on fire....

What's happening out there is a crime so vast it is far beyond a crime to humanity. It is an outrage against all life on this planet. It is a death. It is a murder of the Mother—of the Earth.

I invite you all to witness this. Do it, do it after court hours, go and get the tape which Mark Graham has and which he was not allowed to bring in as evidence.

Hire a helicopter and fly over the area. Witness this, so you can understand. Do it. It is real. Our house is on fire.

Our whole planet is being held hostage by the multinationals. They don't have a conscience because they are not individuals.

The courts are not designed to cope with this challenge. Even the governments are merely pawns in this global game of multinationals. It is only we as individuals, speaking from our collective conscience, who can begin to reclaim our Earth, to turn this thing around.

◆

LUZ MEHER: Your Honor, you think I am naive because I do not think standing on a dirt road to protest insane logging practices is a journey towards the state of anarchy. Maybe you are right, but how can you arrive at chaos via peaceful civil disobedience? To my knowledge no one has ever taken that route before. But that is not all that is uncertain here. For example none of us can say with certainty that my disagreement with the government's decision to cream Clayoquot Sound was not welcomed by society.

We cannot say with certainty that the very society these courts profess to protect did not want us to protest. Nor can we know for certain that society would have granted MacMillan Bloedel the injunction in the first place. Society was not consulted on the matter....

These proceedings have also inspired a significant segment of society to rightly ask what it is that these courts are really protecting. Is it the rule of law that is being threatened or the people who capitalize the most from it...?

Others who did not [blockade]...showed their support for what we did in many other ways. They housed us through the long weeks of this trial, gave us food and drink, they shared their love and sustained us with their compassion.

In fact it was the invisible mass of people who supported the protests that made it all possible. Those people are still out there. They write letters to politicians, to the judicial council of Canada, they plan public forums on the issues they organize in churches, they vote....

At the time of the blockades most of the people I met there wished they could believe that the government sincerely knows what it is doing when it decides an entire ecosystem should be sacrificed for toilet paper and newspapers....

Will these courts demand hungry cougars obey MacMillan Bloedel injunctions when the wild cats are forced to come onto roads and feed on our children? Will your priestly black robes have enough power to demand the tree seedlings adhere to the rule of law when they refuse to grow in harsh and bitter soils? Do you expect to have the luxury of an Appeal Court to right the wrongs you are committing against the laws of nature? If so then I suggest it is you, not I, who is truly naive.

◆

NEIL BURNETT: ...The weight of the law, whose purpose is the maintenance of order and the well-being of the population, is brought to bear upon the "innovators" or "malcontents." The legal arguments employed in the prosecution of these people proceed in almost every case with warnings about the danger of the acts or principles in question to the well-being of the many. And though Justice Bouck draws our attention to the wars

which may have preserved the societies who live ostensibly by these principles, he passes over the struggles, the bloodshed, indeed the deaths of those who were persecuted for challenging within these same societies the existing understanding of the common good and moral order. The Religious Society of Friends (Quakers) has understood this fact for over 300 years.

By the year 1686 there were 1460 Quakers in English prisons (wherein some 450 had died) and many had been hanged in New England. This was precisely because Friends were considered a danger to the social order and legal authority of these places; their convictions and sentences were delivered in tones of grave moral authority. Though the Quakers in these places had never espoused an anarchic position, though they had harmed or defrauded no one, they refused to attend the existing churches, refused to swear oaths, insisted upon allowing women to preach.... This smacked precisely of anarchy to the magistrates.... Who would deny that freedom of worship and the right to affirm in court are essential qualities of this society? The law once rejected them as immoral and anarchic....

This playing of one side against another is one of the most deplorable strategies of MacMillan Bloedel's public relations engine. The stumpage fees amount to a minuscule proportion of the value of the trees taken, and the province spends an enormous amount of money establishing the infrastructure that makes the logging possible. In ten years, most of Clayoquot Sound will be barren clearcut. The loggers will be out of work, salmon streams will have been rendered lifeless, fishing will be harmed and tourism nearly wiped out. The native people will have lost much of their natural and sacred heritage. To allow such an outcome is selfish, short-sighted, morally repugnant and simply foolish.

When faced with the snare laid by his enemies of choosing between loyalty to the state and loyalty to God, Jesus answered, "Whose image do you see on this coin? Then render unto Caesar what is Caesar's and unto God what is God's." It is a saying that Quakers have had constant recourse [to] in their three centuries, whether when assuring authorities that they were law-abiding and tax-paying citizens, in spite of their refusal to fight, or when refusing to give up fugitive slaves to authorities who claimed the

Our planet is dying, our house is on fire....

legal right to seize them. We have seen the forest. It does not bear Caesar's image. After much reflection on what our inner "leadings" might be in response to its destruction, many Victoria Quakers realized that it would be immoral, irreligious, callous and cowardly to stand idly by while the ancient forest was raped and destroyed to the eternal loss of all humankind and all living beings.

◆

JANE FAWKES: ...I am tired of being one of the sheep bleating "Isn't it terrible but it is not my business." Because I am part of the web of life, it is to do with me. I am responsible.

It was for this reason that I stood before the logging trucks on September 23rd. As a woman, I feel empowered by my action. I know I can do something. I am most proud of three events in my life. They were all very difficult. Giving birth to my three sons, riding my bicycle from Vancouver to St. John's Newfoundland, and being arrested at Clayoquot.

◆

TIM WEES: In 1963 I was a participant and quasi-organizer for the Friends of the Student Nonviolent Co-ordinating Committee.... We...attempted to make people aware of the needs of black people in Selma, Alabama. I was for about five years a child care worker...in the centre for emotionally disturbed children.... I ran for parliament in Stormont-Dundas.... I was the editor for...community newspapers...in Thompson and Fort St. John.... In work related to the forest industry, I worked the green chain and then on the assembly line at a stud mill in Taylor, B.C.,...for several years I owned and operated a portable sawmill.... I have driven eighteen-wheeler trucks and currently hold a class 1 licence with air in B.C....

I have heard the criticism that environmentalists at Clayoquot are going to cause a lot of people to lose their jobs. My response is that as the technology of clearcutting is being refined and honed, and as the forests disappear, jobs are already being lost at an alarming rate. In the solution proposed for selective logging, and the rebuilding of a natural forest, and increased local manufacturing, there ultimately will be more jobs and we can realistically rebuild a forest-based economy....

◆

CAROL MACISAAC: Native people begin their meetings with a prayer. I would like to embrace this custom by offering a prayer from the New Brunswick native people: "O Great Spirit, help us to conquer our greatest enemy—ourselves."

I am a music teacher, a Quaker, a mother and a Canadian who loves her land. Although I have great respect for science and for law, I am neither a scientist nor a lawyer. I do not, therefore, presume to fully understand the legal defence to this case, nor did I have the scientific knowledge to convince the court of the necessity of my civil disobedience. (I have been denied the opportunity to bring in the expert witness who could do this.) I do know that I love this land and that I cannot stand by to its wholesale destruction.

I learned to love and respect nature from my family. My maternal grandfather was a gardener. At my family home he planted many beautiful trees that I climbed, played under and watched grow to resemble a forest. I spent many quiet hours sitting in a maple tree that leaned out over a gully. It always calmed me to sit there, listening to the wind rustling the leaves, smelling the natural scents of the bark, earth, leaves and feeling the solid trunk against my back. Most weekends my dad took us out to the country, swimming, hiking, skiing, horseback riding and camping. My paternal grandfather used to take me for early morning swims at the lake. No one else would be up, just us and the lake. My paternal grandmother sent me pictures of trees painted by the Group of Seven for my scrapbook. My maternal grandmother used to take me through the orchard and garden and tell me the names of trees and plants. Since I have left home, my mom would often call me to come up to see the cherry tree in bloom.

All these experiences felt like treasures at the time, but now I see them as spiritual gifts. I hope I can honor these gifts by working to protect a unique rainforest that deserves our protection. I believe my standing on the road at the Kennedy River Bridge was an act of conscience. To quote a fellow Quaker, "The conscience, to Friends, is an expression of God within the individual and is to be followed absolutely. If the promptings are not followed, moral and spiritual sensitivity become calloused" (Robert Byrd, *Quaker Ways in Foreign Policy*).

We are at a point in history where direct and urgent action is needed. If it does not come from government then it must come from people of conscience.

All other avenues available to us have failed: letter writing, phone calls, delegations and demonstrations. Thus the necessity of civil disobedience. I feel privileged to be part of such a peaceful, respectful protest. I have committed no act of violence nor I believe of contempt. Rather, I see my action as a spiritual statement against clearcutting in Clayoquot Sound. Eleanor Roosevelt's words are encouraging: "You must do the things you think you cannot do."

I am not willing to be intimidated by a corrupt corporation and a misguided government in the face of such blatant injustice. I would risk saying that every person in this mass trial has a deep and strong feeling for justice. I cannot say the same for MacMillan Bloedel. At a 1990 blockade on Cortes Island, MacMillan Bloedel stated before a large group that "they can do sustainable logging, but if they did it there, they would have to do it everywhere," and on CBC "Sunday Morning" in August 1993, MacMillan Bloedel admitted that they are cutting as fast as they can because they know they would not be able to log that way much longer.

In October 1993, South American public television was up here filming the "flagrant logging practices and continuous clearcutting." As we are here talking, MacMillan Bloedel is clearcutting our forests. Because of the rate of cutting they are creating a false surplus. They are then forced to export raw logs rather than process the wood here.

We are at a point in history where direct and urgent action is needed. If it does not come from government then it must come from people of conscience. The way they log our land is like farmers eating their seed corn.

In the final analysis, whose land it is anyway? The First Nations of Clayoquot Sound never had a treaty with the Canadian government. Forty percent of this land, their home, has already been destroyed. They are trying to protect what little they have left, a way of life dependant on the forest. The decision by the ombudsman of B.C. (October 1993) recognizes that the aboriginal people hold title to the land in Clayoquot Sound. What of their rights? To quote Elizabeth May, author of *Paradise Won*, a book documenting the struggle to preserve South Moresby: "The least powerful people, the poorest, are those with the most to lose." Who will stand by

them in the fight to save their home and the home of the salmon, the bear and the eagle?

I do not know if my action is justified by the defence of necessity. I simply know that I did what I had to do. I have felt blessed ever since. I believe that the truth sets us free.

To quote Hepzibah Menvhin, the famous pianist, "Freedom means choosing your burden." I choose this burden gladly, with the hope that we will succeed. The fact that ours is the largest act of civil disobedience in Canadian history gives much cause for hope. We have touched the world. As Canadian people we now have the opportunity to gain either the respect or the contempt of the global community. The esteemed Canadian author, Bruce Hutchison, in his book *The Unfinished Country* says:

> "Who will silence the clamor and decide the fate of our species? In the immediate future, perhaps a dozen men at most. But we must look elsewhere for a tolerable life. We must count on the maturing sanity of humankind as a whole."

Another passage further speaks poignantly to my soul:

> "Those who know this land best have breathed the scent of wild rose, cedar, pine, crushed nettle, mossy rock; compressed in sight, sound, scent and in the nation's secret heart are certain memories, regrets and hopes known to us alone. Even if they are far from perfect, we love our native ways and homemade home. There can be no better reason to keep, safeguard and cherish anything of true worth."

...My final words will be another prayer, this time from the Chinook people:

"This prayer I offer to all people and for Clayoquot Sound.

May all we say and all we think,
be in harmony with thee,
God within us,
God beyond us,
maker of the trees."

PUBLIC AND PRESS RESPONSE

The trial judges seemed to take most of the flack from the letter writers whose focus, like most of the protestors, was not that there was an Order of the Court broken, but that there was a Law of Nature being broken. We must take it that any brickbats thrown at the court were really intended for the unholy alliance of MacMillan Bloedel and the provincial government. A sampling follows.

Political satire entered the battle against logging at Vancouver Island's Clayoquot Sound yesterday.

A group called Guerrilla Media distributed 7,000 copies of the four-page *Vancouver Stump* newspaper, placed in newspaper boxes wrapped around copies of the *Vancouver Sun.*

The fake paper's lead story had British Columbia's NDP government merging with MacMillan Bloedel Ltd.

—Canadian Press

◆

The New Democratic Party government is facing the toughest opposition to its decision to allow logging in Clayoquot Sound from its own members, a new Decima poll shows.

—Justine Hunter, *Vancouver Sun*

◆

As a...taxpayer...with an overflowing blue box [for recyclables], I am appalled that the NDP government spent $220,000 to spread propaganda on the logging of Clayoquot Sound....

Our tax dollars are to be spent fortifying the B.C. forest industry's image in Europe—$4.5 million seems an awful lot to spend trying to disguise the lunar landscape.

—John Field, Victoria

◆

On October 26th, 1993, Judge Low stated, "Unless a defendant wishes to apologize, I am not interested in hearing the statements. I've heard it all and I'm tired of hearing about trees."

—The Friends of Clayoquot Sound

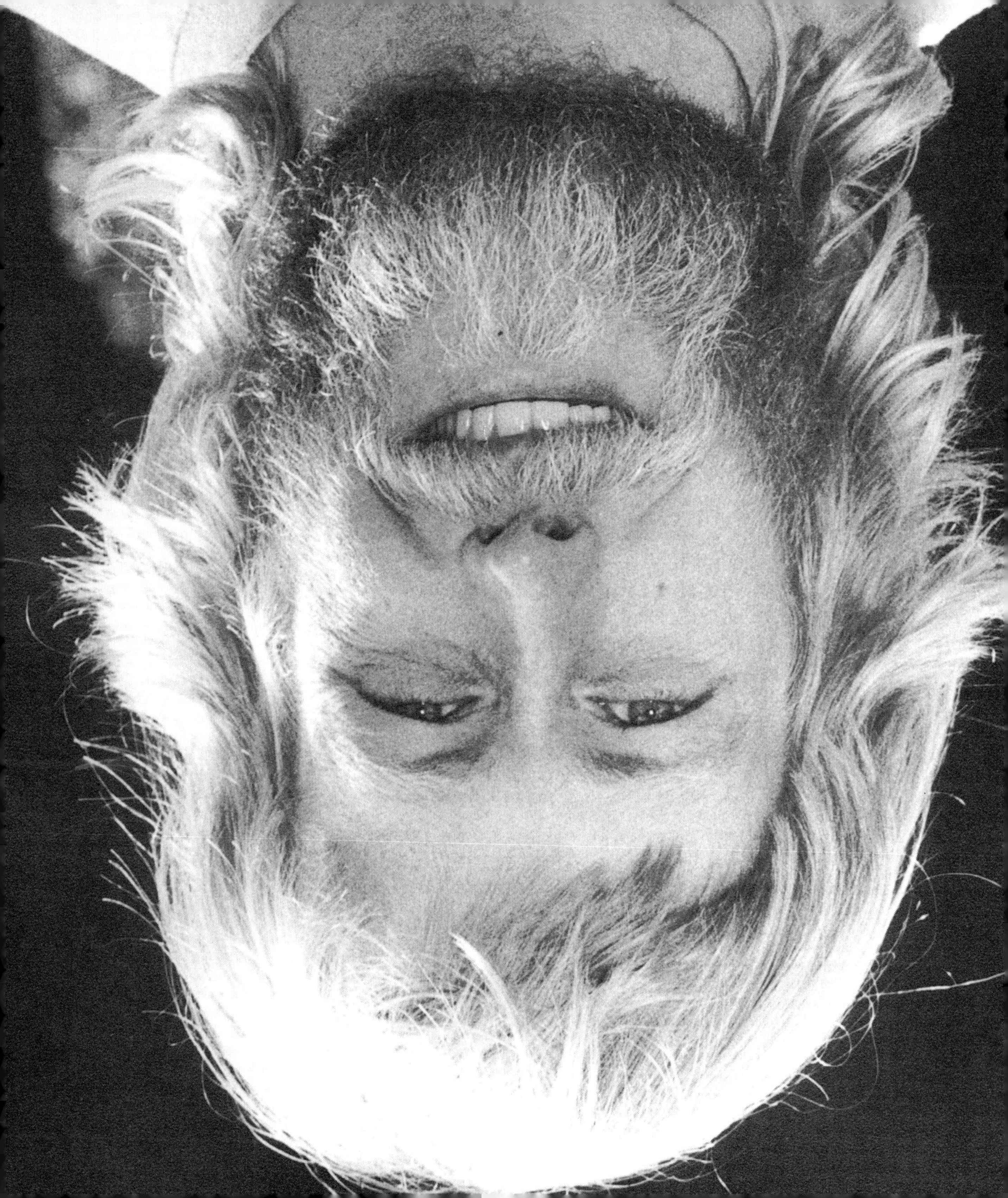

6

SENTENCES AND CELLS: DOING TIME

IF THE CHARGES AGAINST THE Clayoquot protectors drew widespread disbelief and criticism, then the sentences that were meted out by the courts were no less astonishing. Characterized by their inconsistency and seeming arbitrariness as much as their severity, they included 85 days in jail for recalcitrant grandmothers who refused to

stop their protests, and 45 days in jail and a $1,000 fine for the young leader of the B.C. Green Party—equivalent, as he said, to "a second break-and-enter or a first offence for molesting a child." Serving the sentences proved to be as much of an experience for these defenders of nature as the trials themselves, and it remains an open question whether the harsh sentencing strategy will do good or harm in the overall resolution of B.C.'s forest management debacle.

THE SENTENCES

A B.C. Supreme Court justice threw the book yesterday at the first group of 44 Clayoquot logging protestors convicted of contempt of court....

The penalties, substantially more severe than those previously given first-time offenders at similar protests, are necessary, said Bouck, to try to stop the long series of rainforest protests on Vancouver Island.

But some protest supporters vowed the sentences would only increase the number of people to be arrested for contempt of court for ignoring an injunction and blockading the Kennedy Lake logging bridge. More than 700 activists have been arrested since July.

Within five hours of sentencing, Chief Justice Allan McEachern had released on bail half the group who applied to be let out pending an appeal of their convictions.

Protest organizer Tzeporah Berman, who goes on trial in December, said to cheers that the sentences would not deter protests....

Two roving Buddhist monks who have been sitting regularly outside the courthouse say they'll now be taking up a 24-hour-a-day vigil there until the protestors are released.

—Barbara McLintock, *The Province*, October 15, 1993

◆

There has been a wide range of sentences given to date. In the first trial with Justice Bouck, people were sentenced to 45 or 60 days in jail and $1,000-$3,000 fines. Justice Low in the second trial sentenced the majority to 21 days jail or electronic monitoring and $500 fines. One 17-year-old was given a suspended sentence and one year probation. Justice Spencer in the

third trial followed the new precedent, but increased some people's jail time to 28 days and reduced some fines to $250.

The biggest surprise was Justice Murphy in the fourth trial who gave everyone a suspended sentence of 45 days jail time and a $500 fine with a one year probation. Justice Low appeared again at the fifth trial and gave less jail time to those who pleaded guilty to criminal contempt (14 days) and a reduced fine to those who apologized to the court ($250). However, he sentenced one self-represented woman, Carol Johnson, to 28 days electronic monitoring and a $1,000 fine because her approach, in his opinion, was contemptuous of his court.

Since these first trials, some judges have given the option of 25-50 community [service] hours instead of a fine. However, Justice Oliver in the December 8th trial sentenced people with $1,000 and $1,250 fines and no jail time.

The lightest sentence to date is a $250 fine with no jail time given by Justice Drake in the December 17th trial.

—Suzanne Connell, arrestee

◆

Eco-forester Merv Wilkinson and his wife Anne were given suspended sentences Friday for criminal contempt of court, inspiring cheers and applause for the judge from the courtroom of co-offenders and spectators.

Spectators also cheered when Judge Skipp said several mothers of young children would get suspended sentences, a year's probation and 75 hours community work. Skipp said the Crown could justifiably submit that these mothers knew what they were doing. "Nevertheless, I believe that it is time to temper justice with consideration for children," he said. Other offenders got either 21 or 14 days in prison with electronic monitoring urged.

—RFM

◆

THE ISSUE IS CLEARCUT: Anti-logging protester Sheila (Sile) Simpson...got the most severe jail term yet dealt out in the Clayoquot Sound confrontation....

Widespread shock met the severity of her six-month sentence and sympathetic cards, letters, telegrams and phone calls have poured in to her in

Justice Low... sentenced one self-represented woman, Carol Johnson, to 28 days electronic monitoring and a $1,000 fine because her approach, in his opinion, was contemptuous of his court.

prison.

—Greg McIntyre, *The Province*, August 3, 1993

◆

The stiffest sentence [to that point] went to Ron Aspinall of Tofino, who received 60 days in jail and a fine of $3,000. Aspinall was among those granted bail.

—Richard Watts and Roger Stonebanks, *Victoria Times-Colonist*

◆

B.C. Green Party leader Stuart Parker...received a 45-day sentence and a $1,000 fine.

"This is the equivalent of a second break-and-enter or a first offence for molesting a child," said the University of B.C. student.

—Brian Truscott, *The Vancouver Courier*, October 17, 1993

◆

Justice MacDonald...expressed that unfortunately we are forced into a position that if we have a court order, we have to stand behind it. He expressed a concern about labels, criminal and others prejudicial, being applied to well-meaning people. When sentencing defendant Christopher Savery he said "You are not the only one who is concerned," referring to himself and some of his colleagues....

He demonstrated a rare display of respect and individual concern throughout today's proceedings. He decided to go out on a judicial limb and order an expungement of all criminal records relating to this charge for the 20 defendants before him....

After an emotionally overwhelmed defendant, Chris Peterson, expressed that he felt there is nothing else he can do to help save the world from its inevitable destruction, Justice MacDonald sincerely replied, "I am sorry to hear you have given up."

—Clayoquot Sound Resource Centre

◆

The idea that the democracy of Canada has been threatened by the Clayoquot protestors is overstated. If our wonderful country is truly threatened by the action of a few peaceful protestors sitting quietly on a dirt road in

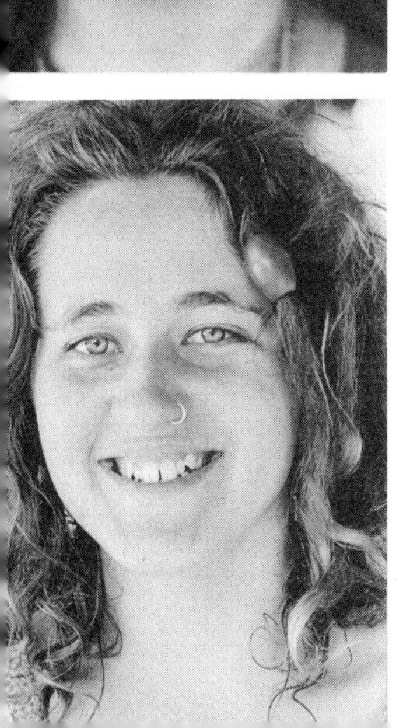

the middle of the night, then it is time to hang up the old hockey stick and move to a different planet. What took place was hardly a "tantrum" but the use of the word suggests the court wishes to condescendingly view the protestors as screaming infants.

In all likelihood, it will have the opposite effect of what was intended. First of all, the severity of the sentence will likely only encourage, not discourage, further action. Second, it will make martyrs of the protesters who have been arrested and convicted. And third, it will further draw the world's attention to the issue and those involved and rightly or wrongly, it is already clear that the world's opinion is strongly on the side of the protestors....

The judgment of history will not likely look favorably upon the reasoning behind the judgment made in the courthouse in Victoria, B.C. October 14, 1993.

—Kim Blank, a faculty member in the English Department at the University of British Columbia

◆

I am writing as an expatriate British Columbian, something which I am most ashamed to admit right now. My small town in Ontario recently raised $2,000 in defence of the old-growth rainforest in Clayoquot Sound.

—Michelle Bull, Port Perry, Ontario

◆

A Salmon Arm psychotherapist offered to counsel Clayoquot Sound loggers and their families as part of her contempt of court sentence.

Chris Morrison, a Clayoquot protestor, offered her professional skills to the logging families.

Morrison suggested that the people who are most likely to need counselling over the dispute are the loggers who are caught in the middle.

—Bruce Patterson, *Victoria Times-Colonist*

◆

Two grandmothers, Judith Robinson (72) and Betty Krawczyk (65)...[have] been imprisoned for 85 days without being convicted of any crime [because they will not undertake to stop their protest].

—Dan Lewis, Friends of Clayoquot Sound publication

DOING TIME

It is still hard to believe that people actually go to jail in Canada in this day and age for protesting in the face of an injunction, or perhaps it is hard to believe that an injunction would be granted in the face of so much evidence of the necessity to protect the planet against overgrazing. Yet very fine people do go to prison for high-minded reasons and the world is a better place for their having gone. Penologists tell me that judges, police and prosecutors should inspect the prisons. Few do, so we are fortunate that we now have many educated and influential people who have prison experience and may work for badly needed change. Here are the observations of one such person:

CLAYOQUOT UNBOUND: A PRISON DIARY

Judge Kenneth MacKenzie shuts his folder. Case closed. A door in the courtroom opens, one we have never noticed before; a door to jail.

MONDAY, JANUARY 10, 1994:

Twenty-one days. Six of us exchange last minute hugs with friends who have sat through two weeks of boring court proceedings. We shoulder our jail bags which include: toilet kit, books, underwear and socks (because those provided in jail itch).

My hat. I am still wearing the Raging Granny hat that started out black and plain the first day of court (in case it should offend the judge) and sprouted flowers and feathers day by day until it now blooms like a Victoria city flower basket. I toss it to the Grannies behind me, feeling absurdly like a departing bride.

"This way, ladies." The woman sheriff's officer sounds patient and totally bored. Six of us file out, all women, too scared to giggle, three choosing jail in a tremulous emulation of Martin Luther King, the others awaiting house arrest on the electronic monitor we all call The Bracelet.

The sheriff's officer puts on rubber gloves before patting us under the arms and down our legs for weapons or drugs and then locks us in a holding cell while someone decides what to do with us. We have a toilet but it is in full view so we try not to use it. We don't even know all each other's names.

Lunch comes, sandwiches and coffee. At 3:00 p.m. the women awaiting electronic monitoring bracelets leave us and we remaining three get the real thing, Smith and Wesson handcuffs. Maryanne Gardias and I are handcuffed together and Susanna Frazer, 75 years old, gets her very own set for the short ride to Wilkinson Road Jail. We are ferried there in a paddy wagon fitted with three cages.

If our wonderful country is truly threatened by the action of a few peaceful protestors sitting quietly on a dirt road in the middle of the night, then it is time to hang up the old hockey stick and move to a different planet.

Wilkinson Road looks so much like a medieval castle that I feel I am in some sort of movie, even more so when I ask politely where we are going and a stolid guard answers: "Segregation: The Hole." Sure I've seen this movie before. Here I am, with not a stitch of clothing on, confronted by a woman guard who tells me to open my mouth, stick out my tongue, shake out my hair, push each ear forward, and then turn around, feet apart, bend down and spread my cheeks. Totally humiliated I comply, and later feel full of admiration for Susanna who refused to do anything of the kind.

In plum-colored sweat suits and men's-size runners, we are marched down the hall to a row of cells nine-foot-square. "Acker! That's yours!" There is a thin mattress on the floor with two grey sheets, a grey blanket, a toilet and a wash basin with taps that do not turn on. The light is bright and fluorescent and there is a slot in the door for the guards to deliver my food. Somebody has scratched a litany of despair on the bars of my window: "Agony, depression, hate, fear, frustration, misery." I try to read the book they allowed me to bring in but Ruth Rendall was not a good choice. Too gruesome. Dinner comes on a plate with some milk and coffee.

Sometime in what seems the middle of the night I press the only switch on the wall. "You have pressed emergency," a voice scolds me. "I thought it was the light switch," I lie. "I control the lights around here," announces the voice. Eventually he does turn them off. I know Susanna will be missing her feather pillow and feather mattress.

TUESDAY, JANUARY 11:

Cereal, toast, coffee for breakfast, pushed through the slot. Wilkinson Road is a men's prison so they will have to send us to either the Burnaby women's unit or to Nanaimo Correctional Centre at Brannon Lake, a mini-

mum security jail up-island where most of the Clayoquot protesters have gone.

In Burnaby we were told the women cons bug you, in Nanaimo it's the men, but they are not "hard core," whatever that means.

We are sent to Nanaimo. This time we are allowed to ride without handcuffs, as are the three young men in the paddy wagon's other cage. It is so strange, watching the sky and trees and then the ocean rush by through the metal mesh.

At Nanaimo Correctional Centre they let us have lunch before processing us: beef, potatoes and chocolate pudding, which would have tasted better if we had not been undergoing inspection by 89 male inmates, some of them muttering, "another bunch of fucking tree-huggers." But there are no complaints about the food or uniforms, forest green shirts and pants, though they seem to come in two sizes, too big and too small. We are allowed to wear our own running shoes, socks and underwear, and to bring in books, writing materials and toiletry. Watches, jewellery and money, which is carefully counted, and the clothes we came in are kept.

There are real mattresses on the beds in our dormitory, plus a table, chairs, TV and a shower. Some earlier protest prisoners have left a stack of books behind with two photographs, one of a clearcut, one of trees to remind us why we are here.

WEDNESDAY, JANUARY 12:

By now we have seen the nurse, the John Howard Society man, the padre and the classifications officer who assigns the jobs at NCC Brannon Lake, which is indeed a work prison and not a holiday camp, he says.

The men, some of them in for two years less a day, work in gangs, making trails at Rathtrevor Beach or testing and repairing fire hoses. Sometimes women get to split wood or work on maintenance but right now the jail is full and work is scarce. There is no work in the winter on the jail farm so Maryanne and I get first shift in the kitchen, which means up at 5:30 a.m., but the shift ends before lunch. Susanna is to be Queen of the Dustballs, in charge of keeping the dorm clean. And we get paid too, twenty-five cents an hour.

Free time from 6 to 8 p.m., we can join the other cons to play pool or ping pong, watch the vicious floor hockey (games that account for a few of the cons on crutches), and check out the hobby shop (where Jimmy Jules is finishing a cedar eagle mask). The library disarray so shocks Susanna, a former school librarian, that she enlists me in the mammoth job of reorganizing the hundreds of Folletts and Ludlums and Micheners and Wallaces. This is a job we never finish and one that may never be finished. On the other hand, it's nice to find cons that read. We are asked for a book on Alberta oil and another about Abraham Lincoln, though its mostly thrillers they want.

"Watch out for drugs hidden there," a guard warns. "What do we do if we do find drugs?" I ask. "Leave them there and tell us quietly about it later." But we never do find drugs, only spiders.

THURSDAY, JANUARY 13:

Just my luck. First morning in the kitchen I meet the deadly meat slicer and lose the tip of my thumb, barely a quarter of an inch, but lots of blood added to the menu's daily protein content. A kind guard takes me to the nursing station, where he stops the bleeding. The jail nurse fixes me up an enormous bandage after breakfast and I am banned from the kitchen.

The men tell me I have lucked into the malingery system with lightning speed, but it is their lightning speed at eating that amazes me. In ten minutes they have demolished beef, corn, mounds of mashed potatoes and chocolate pudding, washed down with milk and are back for seconds before we have properly started. They grumble about the food even as they clean off their plates and threaten to start food fights with left-overs. A man down table from me piles his plate high, pours ketchup and HP sauce on top and bets his neighbour he will eat the lot. There are no takers.

We have been advised not to talk until somebody talks to us. Soon they do. "We're all loggers here, you know. We just love cutting down those trees." One man passes up a large, peeled carrot, "just in case any of you ladies needs one." But their bravado is that of grade school, big kids trying to prove how macho they are.

They look so young—I would say the average age is 23—but they are

"We're all loggers here, you know. We just love cutting down those trees."

not the kids they look. Most have been in jail before, some of them several times. One "kid" fathered his first child when he was 14, he says, and now he has four of them. He is 19 and in for sexual assault. They try to impress us with their wickedness, then insist they are innocent, framed by former buddies who dumped stolen property or drugs on them and ratted. We hear about crooked cops and unfair judges and wives who "ask for a licking." They seem to be the most unsuccessful criminals.

My Vietnamese student would profoundly disagree, of course. I get him by meeting Vivienne, the school teacher, who is trying to coax the brighter, most dedicated inmates through their General Education Certificate while riding herd on the loafers who would rather play computer games or flake out on the couch, scanning a poem or two. But two Vietnamese inmates need help with English, and as they hate each other, they need individual tutors. So Susanna and I take on the job, Tuesdays and Thursdays.

My student (I will call him Lao) does not really care if no one understands him. He talks anyway, in a wild mixture of consonants. He understands English very well and Chinese and Thai, and boasts delightedly about the millions he used to make in Macao and Hong Kong where business was so much better than B.C., where we are all lazy, he says. He has a late model luxury sports car, a bullet in the neck from trouble with another gang, and a passion for violent movies with lots of explosions.

"Lao bad boy," he tells me happily. I ask no questions. But the compositions he dictates and then learns to pronounce are almost as good as a Ludlum thriller. Then he tells me he is very lonely, that he telephones his son every night and he is very sad. He is the only one who says what most of them must be feeling, especially those in for a longer sentence. No wonder they look at us a little wryly, idiots who chose to go to jail.

FRIDAY, JANUARY 14:

Mail. It's delivered opened in case it includes drugs or maybe getaway plans or maps to show where the loot is stashed. But it's great to receive cards from friends and supporters, postcards of Clayoquot or of any old tree, because we cannot see any from our window.

But that was before we got the summons to move house, from the dorm

where we had to press a button to enter the hall and identify ourselves: "Female inmates to the dining hall." Now we were headed for House No. 1, formerly a staff house close to the lake. We have to go through two locked gates in a 12-foot fence to get there. The chief security element, however, will be close beside us in his kennel: Fluffy, the Rottweiler guard dog.

"I wouldn't try to make friends if I were you," we are warned. "And if he's loose, keep the door shut tight."

Fluffy and his handler patrol the grounds ten hours a night. When we meet them both on our way back from dinner, he lunges, teeth bared, growling convincingly.

We disturb him at first when he's trying to sleep in the mornings and his handler accuses us of tiring him out, but we soon get used to each other, though we were never chums. There is a kitchen cat called Marmalade and an aquarium in the school room for Stan the fish. We "catnap" Marmalade one day but return him to the kitchen. He knows where the best eats are.

In our new house we have three bedrooms, a kitchen, a bath with a shower, TV and a view of the lake where a flight of geese settles. There are cows across the lake and a real dog—a free dog—barks happily. We do Tai Chi on the patch of lawn where we cannot disturb Fluffy's sleep.

SATURDAY, JANUARY 15:

Visiting day. There are two visitors for me, both nervous about signs that warn them not to dress provocatively or bring in food or drugs. They can buy me candy at the canteen. There is coffee laid on in Park House, and toys for the kids to play with. The con who sits beside me in the dining room crouches down to hug his three daughters and fetches the youngest a crimson teddy bear. His wife brings them every Saturday. Another woman has brought three red roses "for our anniversary." It's a precious time for us all.

SUNDAY, JANUARY 16:

Church: Reverend McLusky has brought in young people from the Thetis Island Bible College. Timo, from Germany, plays the out-of-tune piano, Irene sings, and Del plays the flute and tells us a gloomy story about death

Six of us file out, all women, too scared to giggle, three choosing jail in a tremulous emulation of Martin Luther King, the others awaiting house arrest on the electronic monitor we all call The Bracelet.

and sin. The cons wander in and out or sit on the edges of their chairs, some trying to stop the twitch that comes with drug withdrawal.

"We don't expect too much," the Reverend admits," but they like to know someone cares."

MONDAY, JANUARY 17:

Ten more women protestors arrive. Five should have gone on the bracelet home monitoring program in Victoria and are here by some bureaucratic error. They will be returned and confined in their homes on Tuesday, but another protestor, a man, spends a week among the sex offenders before the red tape catches up with him. He pronounces his stay interesting and uneventful.

With eight women now in our house we have to set our own rules: no morning showers or there is a bathroom traffic jam. Smokers go outside and we all clear up after ourselves. We tell each other our life histories, exchange trial gossip and plan not to waste our time watching TV or sleeping. There is Tai Chi outside every morning and talk circles in the evening.

One newcomer, a professional musician, is allowed her guitar. She wakes us up with Good Morning, Star Shine. I learn the medicine cards and pick my sacred animal, an elk. I would have preferred an eagle or a mountain lion but the elk is patient and wise.

The younger ones of us are heavily into crystals and the Tarot and energy forces. I am a sceptic but find myself absorbing their peace and trust. They are a brave and hopeful group. They make bead necklaces, paint butterflies, write poems, tell me their adventures hopping freight trains and announce, most of them, that they do not want their parents to know they are in jail because they would not understand. I would probably feel the same way if I was 19. They have great plans for changing the world and take heart from each other. We do a lot of hugging.

Sometimes it is hard for them to remember that the cons are not always the nice guys they seem, and that the rules about not being alone with a con are based on good sense. They get dates for ping pong and one young women receives a love letter. "If I were free, I would bring you a rose." A guard tells me about an earlier protestor who burst into tears when he had

to tell her twice to tuck her shirt into her pants, wailing, "Nobody ever yelled at me before."

"We have had a few lost souls, but most of you have been a good influence on the men. We are proud to be the only co-ed jail in B.C. and we want to keep our good record."

I am two-minded at turning into such a co-operative con, but try to walk the straight line, not talking too much to the cons or the guards, doing my time cheerfully. If any cons want to talk trees, that's fine. But I would rather listen to their ideas, which are as varied as those outside, once they get beyond the initial claim that every one of them is a logger. It goes with the image.

TUESDAY, WEDNESDAY, THURSDAY, FRIDAY, JANUARY 18-21:

Time rushes now, partly because we get a real job, breaking down basement walls with crowbars so that the new rooms can be built downstairs. They must trust us. What a joy to smash beams and pull out nails. I am unnerved by my own powers of destruction, and somewhat relieved to find one of us is upset by spiders.

We are enjoying ourselves so much we get the work done too fast, and have to ration ourselves to mopping half the floor one day and half the next.

Work helps my growing sense of frustration at the arbitrariness of rules. Some guards let us walk home without an escort, or wait in the school room for twenty minutes without supervision. Others, usually the younger guards, stick to the rules and give themselves more work to do. Ma Miller, the second in command, seems to achieve the most by her consistency and gruff friendliness. She has been in the business for 14 years. We send her a thank you letter when we leave.

SATURDAY, JANUARY 22:

Our last day. I have been lazy and I have eaten too much but I hope I have used my time wisely. I have certainly learned a lot, and have come to respect both the guards and the cons, who are all prisoners of a system that

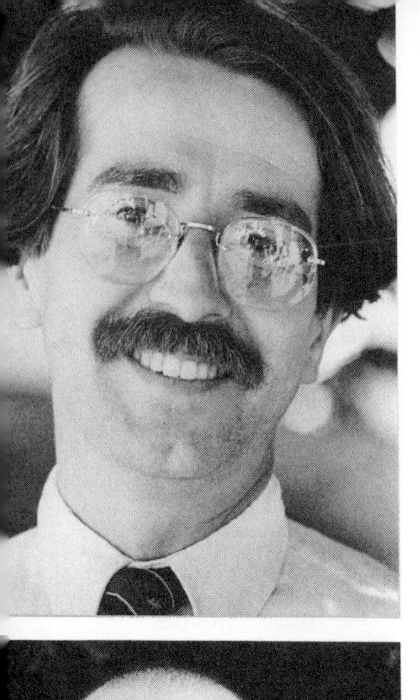

does not make much sense and are trying to make the best of it. That goes for Fluffy too. I resolve to send him a card when I get out and I do.

SUNDAY, JANUARY 23:

Off with the forest green uniforms and into our own clothes. Our valuables are returned, everything carefully noted. I am $40 richer from my 25-cents-an-hour, which I put towards Jimmy Jules' eagle mask.

"Your friends outside are making quite a noise," the guard reports with some concern. The Raging Grannies are there, of course. I learn later that they had given him a hard time for referring to us as girls instead of women.

I am no longer Female Inmate 8304620. Was it worth it, going to jail for a principle? It was worth it to me. Jail was neither a picnic nor the Bastille. I do not feel like a heroine, but I have not repented either. Cons and guards, loggers and protestors—we are all in this together. Sooner or later we have got to find a better way out of the woods than fighting each other and going to jail.

Tannebaum Revisited

The jails are filling up with thieves,
With murderers and muggers,
But deadliest of all of these,
Are Clayoquot tree-huggers.

They block the road, they sing in jail,
So badly that the warders quail,
Some say they don't just hug the trees,
Some of them hug loggers.

Home on the Range

If you have to serve time for the terrible crime
Of protesting the clearcut of trees,
Let's hope you go to jail where there is male and female,
And the guards are as decent as these.

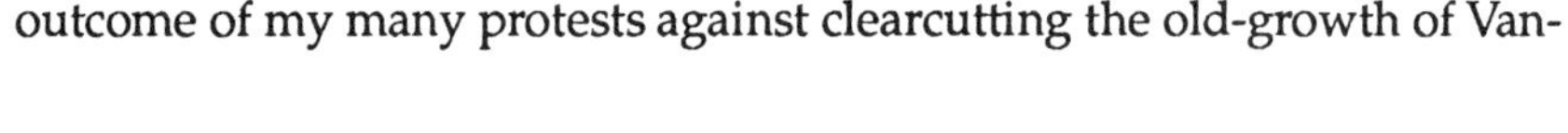

CHORUS
Brannon Lake, where they serve you pizza and cake,
If you play by the rules,
And you don't act like fools,
It's not such a hard place to take.

—Alison Acker, a writer with three books published; a retired English professor from Ryerson Polytechnical Institute, Toronto; has three grandchildren. Her recent history is exciting: she spent January-April 1993 in the Guatemalan jungle as an accompanier protecting 2,500 returning refugees from the army and the death squads, as a member of Project Accompaniment. She is a member of the Raging Grannies.

◆

Doctor Kleiman kept careful notes that indicate changes needed in the jail system. Her Quaker group is active in prison visitation and in the search for justice behind the walls. Here is a summary of her story:

PRISON AT BRANNON LAKE

I carried my black tapestry bag with my art supplies and books to the courthouse for three days, because no one could predict when my trial would end and I would be immediately imprisoned. It felt like being pregnant; I never knew when my water would break and begin the events that would irrevocably change my life.

Finally, Judge Hutchison said,"Twenty-one days and $250, the fine to be translated into 25 hours of community work."

Throughout the trial, the prosecuting attorney and Judge Hutchison reassured us they understood we were not criminals. In a symbolic gesture at the Clayoquot blockade, I had delayed a multinational's logging trucks for less than two minutes for no personal gain, but to attempt to insure old-growth forests would be available to the children of the future. In the end we received a harsher sentence than a man who for years sexually molested his step-daughter, an interesting commentary on our society's values.

I could have chosen electronic monitoring, but chose jail as the logical outcome of my many protests against clearcutting the old-growth of Van-

couver Island. As a Quaker, I would bear witness to this wrongdoing by going to jail....

We were very pleased to be at Brannon Lake near Nanaimo on Vancouver Island, because three of the four of us were mothers with children who would visit. As we stood nervously in line for our first meal, a guard passed behind us and muttered, "You think you're so smart. You won't last two days in here. We'll get rid of you."

The corrections officers were mainly friendly. Several thanked us for our courage in going to jail. Two had been up to participate in the blockade, but could not be arrested because of their jobs. Two men Protectors were repeatedly physically threatened in their dormitory.

My language coarsened as I learned to protect myself. My job was splitting cedar kindling for the Boy Scouts Christmas sale. (Ironically, my son Jake had sold these kindling bundles when we lived on Gabriola Island....)

The television was usually playing MuchMusic at disco decibels.... We saw such videos as "Live from Death Row." Most films were saturated with violence against women, which saddened and angered us.

The rage and hostility of the other prisoners was always just at or below the surface. Emotions like boredom, loneliness and grief predominated. In prison, turning your emotion off is the only way to survive. You withdraw into yourself and "do time." The prisoners harassed each other and sexually harassed us. One...suggested they should build us padded benches so we would be comfortable when we performed oral sex on them all. I am a child of the '60s but in prison was offered more drugs than I had ever seen in the '60s....

One CO told me they would be happy when marijuana became legal, because the men seemed to "mellow out" when they smoked pot. When they drank the illegal alcohol, there were knife fights and people were badly hurt.... The long hours of boredom were always tinged with tension and fear....

When we were released from jail, we scattered to the winds. I went home, but was profoundly affected by my experience. I visited the Clayoquot, but returned still sad and unsettled. Marty Heiken, a man who did not know me personally, lent me his family's Pender Island cabin. For a

Cons and guards, loggers and protestors—we are all in this together. Sooner or later we have got to find a better way out of the woods than fighting each other and going to jail.

week I stayed there writing a pamphlet about refugee mental health for my community hours. I walked in the woods every day with my dog. Slowly I healed and then returned to work.

—Betty (Crystal) Kleiman, a 54-year-old Victoria mother, pediatrician and artist. She grew up in New York City and never saw a forest until in her late teens. She has lived in British Columbia, Canada for 25 years and enjoys kayaking and camping in the old-growth forests near her home on Vancouver Island.

◆

The leader of the B.C. Green Party wrote this letter from his jail cell:

...Political opposition to the vested interests that rape our land and poison our waters and air will not be tolerated. Is it any wonder we have a Third World in which we are called the "Brazil of the North"? It is not just the grapple-yarders and burning forests, it is the smiling, glad-handing politicians who try to mask the clearcut scars and dying forests, the mass trials, the shackles and handcuffs, with empty words about new forest practices and public input. With their massive ad campaigns, their well-placed political donations, their commissioned polls, their tactics of intimidation and fear, the transnational forest companies have controlled this province's governments, economy and people for generations.

Staring out of my window at the razor wire enclosures, I wonder if things can change. I hope that they can, that people can count how many advertising dollars the forest industry spends for every cent spent by Greenpeace, that people will go to the forest themselves and see the destruction and loss, that people will remember the East Coast fishery was also called a renewable resource and that, for my generation and our children, they stop the madness that passes for forestry in this province.

—Stuart Parker

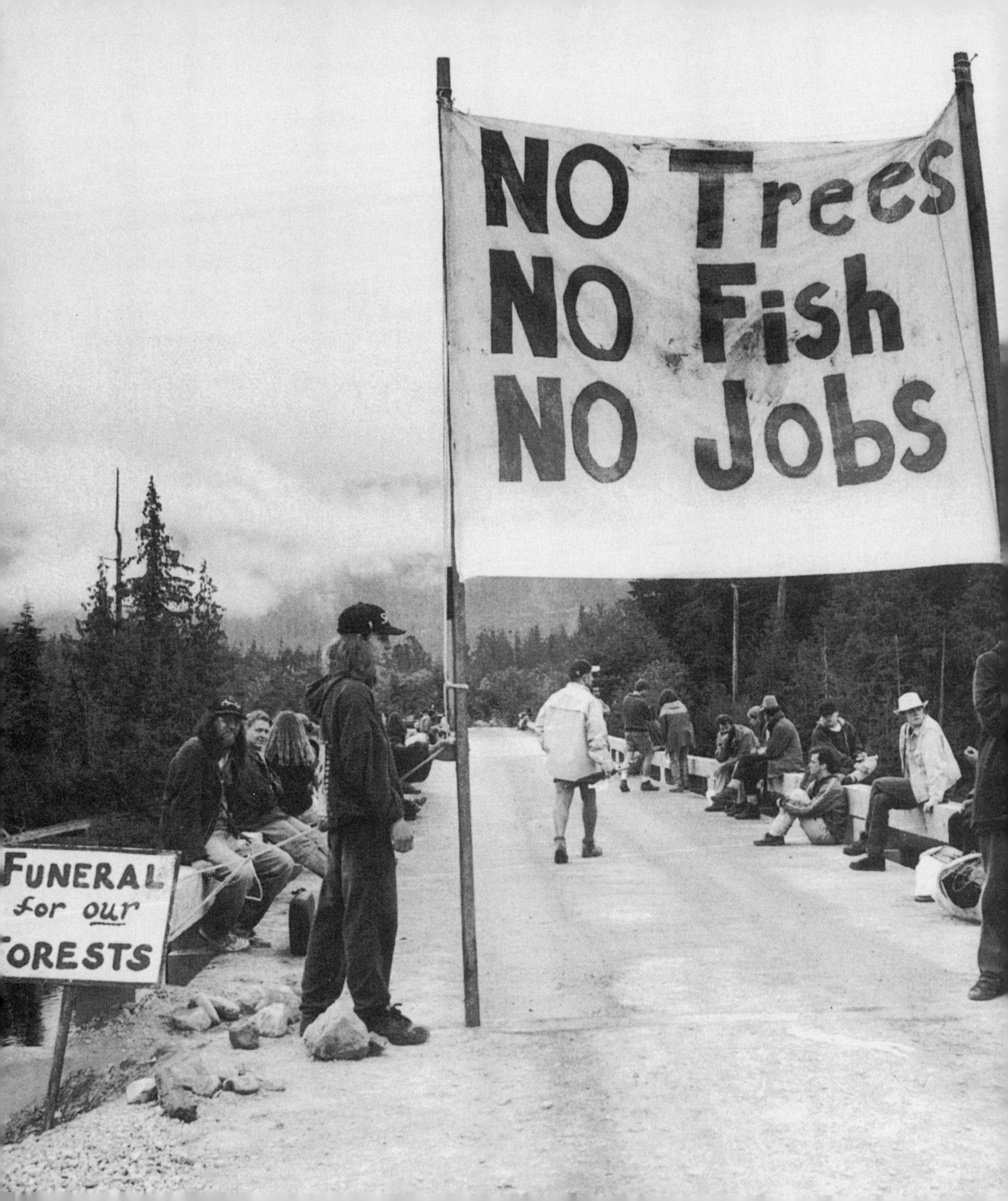

NO Trees
NO Fish
NO Jobs
FUNERAL
for our

7

THE FUTURE: A REPLAY OF THE PAST?

IN THE WAKE OF THE ARRESTS at Clayoquot and the ensuing charade in the courts, the B.C. government insisted that it would not bend even the slightest bit on its position concerning forestry in the Sound. The repercussions of both the protests and the government's intransigence went far and wide, however, and the protectors—far

"It's time to go into the next stage, it's time to start organizing a boycott of B.C. products. They are not going to listen to anything else."

from abandoning the struggle—reinforced their international campaign, attempting to persuade overseas buyers of Clayoquot wood to cancel their contracts with the offending multi-national forest companies.

Back in Canada, meanwhile, the question was still not whether it was possible to do forestry in an ecologically-sound manner, since almost everyone now knows that it is, but whether the protests had finally created the political will to insist that this was to be the way forward.

◆

Prime Minister-designate Jean Chrétien says he is prepared to talk to the B.C. government about converting Clayoquot Sound into a national park but Premier Harcourt dismissed the park notion Wednesday as laughable.

Chrétien avoided the critical question, "Who pays?" Lost jobs, timber rights and natives would claim millions. Chris Lee, leader of the Canadian Green Party, said outside the courthouse the ruling [against Clayoquot protestors] meant it was time for environmentalists to take tough action. "It's time to go into the next stage, it's time to start organizing a boycott of B.C. products. They are not going to listen to anything else," he said.

Dave Haggard, president of the International Woodworkers of America Local 1-85, said the protest had put such an emotional strain on the loggers of Ucluelet that the union has to hire two counsellors to help them deal with the stress.

Ucluelet children are watching the protestors on television and starting to question what their parents are doing. "I have seen families torn apart over the last six months," said Haggard.

Logging officials have estimated that protecting Clayoquot Sound would eliminate 1,200 jobs on Vancouver Island and cost $300 million in lost economic activity.

Environmentalists say those jobs will disappear anyway once the timber is gone. They say protecting the Sound would bring more benefits in the long run through tourism.

—Victoria Times-Colonist

◆

Whether our effort will meet with success remains to be seen. If not the failure will not belong to the protestors but to those who refuse to acknow-

ledge the wisdom of preserving that which maintains life in ways we have not even begun to understand.

—**Jean Bready Randall, 52, arrested August 9, 1993**

◆

The Tofino-based environmental group [Friends of Clayoquot Sound] is now switching its focus to MacMillan Bloedel's customers. Environmentalists recently flew to California where they met newspaper companies and other buyers of pulp and paper. The environmentalists told the companies that by purchasing goods from MacBlo they are contributing to the destruction of Clayoquot Sound, the last, largest unbroken expanse of temperate rainforest left on Vancouver Island.

—Richard Watts, *Victoria Times-Colonist*

◆

HOW TO REGAIN CONTROL OF OUR FORESTS: To eliminate the current forest tenures, all the provincial government needs to do is pass a law cancelling them. The question then is "how much compensation is due?"... The big companies paid only token fees for their licences and have already realized extraordinary profits on their initial investment. It makes much more sense...that they be required to buy the wood they need for their mills on an open B.C. log market....

The provincial government should...expropriate Tree Farm Licences...[and]:

(1) Outlaw clearcut logging and require selection harvesting systems. Require minimum employment commitment per cubic metre of trees cut. Require...wood get sold to the highest B.C. bidder. This will encourage value-added manufacturing, giving many entrepreneurs an access to wood they are now denied. The large companies will claim that it is impossible to do this; that they cannot afford to make these improvements and still remain competitive on the world market. If they insist on producing low value wood chips, pulp and two-by-fours, they're probably right. Their choice will either be to change and meet the new rules—which will create more jobs, use the wood more fully and sustain the forests—or abandon their holdings to companies that will.

(2) Formally recognize aboriginal title and proceed to invalidate the Tree

Farm Licences based on the natives' prior claim on the land. Some of the licensed land must go to First Nations to give them the land base they need to support their communities. Some must become much needed wilderness preserves. The remainder must be converted to community-managed forest reserves.

(3) [Make] the big companies pay fair taxes.... They have taken advantage of every tax break and government subsidy imaginable.

—*A Conservation Vision for Vancouver Island*, **Western Canada Wilderness Committee Educational Report, Winter 1993-4**

◆

In China, 75 percent of their 11 million tonne paper output is produced without using tree pulp.

—**James MacKinnon,** ***Monday Magazine***

◆

It...borders on criminal that our government and the multinational logging companies have not created hundreds of jobs in rehabilitation of streams, bogs, and other ecological disasters they have created.

—**Cameron Delacroix, Victoria**

◆

MacBlo is sending loggers to school for a whole day to teach them not to mess up the streams, rivers and fish habitat...I don't know whether to laugh or cry.

—**Martin Hykin, Victoria**

◆

When environmentalists fight loggers, it's not just over trees. Traditional logging roads that move equipment and carry away the bounty leave zigzag scars on once virgin slopes. By using helicopters instead of trucks, the logging and land management firm Champion International has Washington State preservationists joyful and hikers cautiously optimistic. The company had three reasons for choosing the rare helicopter method to clear 160 trees off Cedar Butte: the area is highly visible from Interstate 90, it is directly above a State park, and there is a popular hiking trail. Since last month's logging, Champion has received queries from logging operations

"What's being done in B.C. is very tragic. It's very sad and it's bad business."

and praise from environmentalists.

—**RFM**

◆

LONG-LINE MACHINES EAGERLY SOUGHT: Bob Norcross spent much of his...life operating long-line logging rigs.... They require fewer roads and employ more men.... He saw the old systems based on spars and towers being replaced by grapple-yarders—as ugly and as destructive to the land as their name implies....

Skyline logging—hauling logs off the land with a suspended long line—is the way of the old-timers.... Skylines are environmentally more friendly as no heavy equipment is on the fragile forest floor. The system also benefits the company.

"There's no breakage. The logs are in perfect bucking condition when they get to the landing," says Dave Summers, superintendent of Woss Lake operations for Canfor Corp.

—**Gordon Hamilton, *Vancouver Sun*, October 19, 1993**

◆

TOP JAPANESE EXECUTIVE SHOCKED BY PHOTOS OF B.C. LOGGING ABUSES: Iwao Terui, Associate Vice-President and Senior Executive Manager for the Telephone Directory Division of Nippon Telegraph and Telephone (NTT), expressed shock today upon viewing photographs of MacMillan Bloedel (MB) logging practices in the region of Port Alberni, B.C. MB is supplying 60,000 tons of paper for NTT's annual telephone directories. Paper for the three-year contract comes from MB"s Port Alberni mill, adjacent to Clayoquot Sound. After seeing photographs documenting MB's clearcutting in the region, Terui said, "I just can't imagine our telephone books being printed on those beautiful trees. What's being done in B.C. is very tragic. It's very sad and it's bad business."

—**Friends of Clayoquot Sound News Release, November 22, 1993**

◆

MB BULLIES CRY THE BLUES: In the wake of Scott Paper UK's major contract cancellation, MacMillan Bloedel has taken to flooding all media with an advertising plea for Canadians to stand up against intimidation and lies from the environmental movement.

MacMillan Bloedel has taken to flooding all media with an advertising plea for Canadians to stand up against intimidation and lies from the environmental movement.

At a recent seminar in Brussels, MB Woodlands manager Bill Cafferata admitted that 25 percent of the ancient temperate rainforests they manage are being clearcut for pulp. ...Friends of Clayoquot Sound activists have presented briefs at the European parliament and are meeting with lumber clients of MB in Europe as well as giving slideshows and talks.

—Greenpeace News Release, March 14, 1994

◆

MB LOST IN THE WOODS AGAIN: MacMillan Bloedel has admitted today to mistakenly logging in an 11.7 acre area in Clayoquot Sound early last month. The reason given was that the fallers are not used to working in cut blocks this small and cold not locate themselves on the map. Two and a half acres of pristine forest was felled before the error was discovered.

—Friends of Clayoquot Sound News Release, March 4, 1994

◆

Tony Duggleby is a handlogger who operates under the Small Business Forest Enterprise Program on the coast. His company, Eclectic Logging, generates three times as many jobs, and generally seven—but even as high as 15—times the revenue generated per cubic metre of wood harvested by the major forest companies. Additionally, the SBFEP loggers pay, on average, twice as much as the big forest companies in stumpage on Vancouver Island.

—*A Conservation Vision for Vancouver Island,* Western Canada Wilderness Committee Educational Report Winter 1993-94

◆

Fritz Leutgeb, the founder of an Armstrong value-added manufacturer,...manufactures laminated window components for export to Germany, Austria and Switzerland.... [He] said that if they could find the timber, there is enough demand to keep ten factories like his busy.... [Vernon forest district's small business forester] Jim Smith explained, "This province could be extremely wealthy if timber was sold competitively."

—Jim Cooperman, *British Columbia Environmental Report,* October 1993

◆

POLL SHOWS BRITISH COLUMBIANS WANT LESS CLEARCUTTING—Environmental groups file formal complaint against *Vancouver Sun*: A coalition of major B.C. environmental groups today released the full version of a poll commissioned by the B.C. government that shows the B.C. public want dramatic changes in provincial forestry practices. Yesterday the *Vancouver Sun* published a story about the same poll but failed to report many of the key findings and misrepresented or manufactured others.

"This poll shows strong support for the platforms put forward by environmental groups. A total of 93 percent think clearcutting should be banned or reduced in size," said Greenpeace Forest campaigner Karen Mahon.

—Greenpeace News Release, May 11, 1994

◆

BUSINESS AS USUAL IN CLAYOQUOT SOUND: *An Independent Assessment of Active and Recent Cutblocks in Clayoquot Sound and the Port Alberni Forest District (prepared by the Sierra Legal Defense Fund, July 1994).*

An audit on active and recent clearcutting in Clayoquot Sound, the Tsitika and other locations on Vancouver Island shows that logging companies continue to violate the B.C. Fisheries/Forestry Guidelines and destroy wild salmon habitat.

Seven of the cutblocks examined were operations by MacMillan Bloedel in Clayoquot Sound, an area where the highest standards have been promised by both the B.C. government and the company. The field studies reveal that the logging giant's "new forestry" fails to meet even the most basic standards.

The report is based on field work carried out in the last two months by Sierra Legal Defence Fund's environmental investigations team which consists of a registered professional biologist and a lawyer specializing in environmental law.

Ten randomly-chosen cutblocks, all in proximity to fish-bearing waters, were inspected. There was at least one minor or major infraction of the Guidelines on every single cutblock visited, and over 60 probable violations of assorted rules, guidelines and permits. ..."Every active clearcut we studied had problems, " said John Werring, Registered Professional Biologist with the Sierra Legal Defence Fund. "Logging debris clogged streams,

Field studies reveal that the logging giant's "new forestry" fails to meet even the most basic standards.

there was improper drainage on roads and inadequate leave strips, even based on existing guidelines."

According to research by the Friends of Clayoquot Sound, since the announcement of several new government initiatives, including the Central Region Board which includes the Nuu-chah-nulth First Nations, over 48 new permits for logging and roading in Clayoquot Sound have been issued. These permits fly in the face of recommendations made in interim reports of the Scientific Panel for Sustainable Forest Practices in Clayoquot Sound.

Greenpeace is demanding that the Central Region Board approve all cut permits, and that the B.C. government adopt the new American PACFISH guidelines for riparian protection zones as a minimum interim standard to protect fish habitat. The U.S. has decided to protect Pacific Northwest salmon habitat on Federal land with 90-metre leave strips along any stream bearing any kind of fish. The U.S. does not permit harvesting within the leave area.

— Delores Broten, *Watershed Sentinel,* **August-November 1994**

◆

B.C.'S NEW FOREST PRACTICES CODE DOES NOT MEASURE UP TO U.S. STANDARDS: The Sierra Legal Defence Fund (SLDF) has teamed up with the Natural Resources Defense Council to produce a report comparing the regulation of forest practices in British Columbia and Washington State.

...We looked at ten aspects of forestry regulation, and found that the U.S. forestry laws are generally much stronger than the B.C. Forest Practices Code. U.S. laws tend to prescribe agency action, rather than just enable it.

...Despite the government's promises and promotional material, B.C.'s forest policies do not appear to be "world class," especially when compared to the policies south of the border.

—Mark Haddock, Sierra Legal Defence Fund, in *British Columbia Environmental Report,* **September 1994**

◆

"WORLD CLASS" CODE PR SCAM—MAJOR B.C. GROUPS BLAST CODE: The long-awaited release of the Standards and Regulations which get to the heart of forest practices in B.C. was greeted with universal dis-

may by the province's major environmental groups and by grassroots organizations alike.

Meanwhile, the Harcourt government continued its attacks on "fringe" environmental groups, in an attempt to drown out criticism in Europe and the provincial press.

...Criticism of the new Forest Practices Code by Western Canada Wilderness Committee, Greenpeace, Friends of Clayoquot Sound and the Valhalla Society is shared by many other environmental groups.

"I resent Harcourt's attempt to dismiss criticisms of the code by labelling some groups 'fringe'," said Greg McDade, executive director of Sierra Legal Defence Fund. "Most major environmental organizations have real problems with the code because there appears to be an escape clause for every regulation and because too much discretionary power rests with district forest managers. It's basically a status quo document that won't be tough enough to correct the problems in the forest."

—Delores Broten, *Watershed Sentinel*, **August-November 1994**

◆

APPEALS DISMISSED: The January 1994 appeal by British Columbia's Sierra Legal Defence Fund against MacMillan Bloedel's injunction in Clayoquot Sound (*Greenpeace Canada and Langer v. MacMillan Bloedel Limited*) was finally decided on September 30. By a two to one majority, the Court of Appeal dismissed the challenge to the injunction. The decision will be appealed to the Supreme Court of Canada.

The actual conviction appeals (from the first 45-person trial) were argued before the Court of Appeal at the same time by a group of criminal lawyers. Those appeals were dismissed by the Court of Appeal on March 28, 1994. Those convicted then applied for Leave to Appeal with the Supreme Court of Canada. On October 6, 1994, the Supreme Court dismissed those applications for Leave, with one exception. Leave to Appeal was granted to one individual (Jonathan Pulker) concerning the issue of whether the Young Offenders Act applied to preclude a conempt of court conviction.

—Greg McDade, Sierra Legal Defence Fund, October 7, 1994

PROTEC
CLAYOQU

EPILOGUE

Valerie Langer

Thomas Jefferson said, "The price of freedom is constant vigilance." What is to be said for a government and court in a democratic nation which jails its most peaceful, most democracy-loving, most vigilant citizens—its exemplary citizens? In ordering the mass trials to proceed, British Columbia's Harcourt government and the B.C. Supreme Court positioned themselves in opposition to altruism, peacefulness, environmentalism, eco-foresters, youth, Roosevelt Elk, marbled murrelets, and wild salmon. It was the disenfranchisement of those who had, up until the Clayoquot Blockades, considered themselves full members of Canadian society.

The B.C. government misjudged the Canadian public's capacity for uprising and its committment to Clayoquot. Then, through the media and its instructions to the Crown Counsel, the government insulted those who demonstrated and those who supported them. Finally, they persecuted intelligent, peaceful, articulate citizens. These arrogant, dissmissive gestures helped to solidify the progressive movement. The mass trials were yet another indication of a government and judiciary out of touch with the pulse of the people.

By the time the first 44 were sentenced, Canada had already been treated to images of both beauty and clearcut devastation in Clayoquot Sound. Canadians found out that they had a rainforest in their country at the same time as they found out that it was being destroyed at an alarming rate. Their televisions barraged them with images of hundreds of people like themselves being carried off the blockades by police. They heard interviews on CBC, campus radio, and Country music stations about how the Peace Camp functioned on principles of nonviolence. They saw banners which read "Forest Industry Workers Against Non-Sustainable Logging;" placards saying "NO FISH, NO TREES, NO JOBS;" articles which decried the brutal logging practices. Then their cousins, friends, grandparents, priests or co-workers drove out and participated in the blockade. By the time the first 44 were sent to jail, Clayoquot had caught the public imagination.

The sentences sent a shock wave across the nation and beyond our borders. Perhaps for the first time in their lives, people felt personally affected by the jailings of political activists. These were not activists in China or from some exotic, far-removed dictatorship; they were not fringe anarchists, nor native peoples. The convicted were people with lives like their own who believed in the same things they believe in. With that first shock wave came a wave of cynicism about our justice system. Comfortable Canada suddenly faced oppression when its citizens challenged the status quo. Friends were being arrested, silenced, jailed. The sort of politicization that resulted is the kind that no group can organize. No pamphlets or educational video could have done for the grassroots environmental movement what the Supreme Court of Canada and the B.C. government did. It's an irony they most likely do not appreciate!

With the mass trials, thousands more people became active in the Clayoquot Sound/temperate rainforest movement from a democratic justice point of view. If they were only marginally involved with the forests issue before, the democratic justice challenge spurred them into becoming activists. The letters to the editors, the speeches, the campaign took on a new tone raising the temperate rainforest campaign to being part of one of the most important movements in history—the movement against the extinction of species. The fact that such a vociferous, powerful, grassroots movement has built in Canada is all the more surprising given the soft-stepping Canadian culture.

There are some things one cannot predict nor manufacture. Nobody manufactured the Clayoquot movement; it developed in an incredible, bounding manner. It became a symbol. With every tyrannical attempt to silence the movement, the government challenged what Canadians deeply believe to be their rights and took a swipe at some of our basic cultural assumptions: the right to be treated justly under the law, the right to protest, our expectations of honesty, our loathing of corruption, and the assumption that we live in a democratic country blessed with a beautiful and vast wilderness.

As a campaigner for Clayoquot Sound and the temperate rainforests of

British Columbia, I couldn't help but feel that the government's heavy-handed approach to the Clayoquot arrestees and their harsh sentencing made for very good campaign material. As a citizen, an activist, an altruist, I have been greatly saddened by the mass trials. My heart is with the 857 who put their liberty on the line for the forests. Their courage and dedication inspired a nation and made them heroes of the grassroots environmental movement. Yes, it happened in our "Canada-the-good." As we continue to challenge the destruction of forest ecosystems, we will continue to face the heavy hand of the status quo's proponents. — That's where the "constant vigilance" comes in....

PROTECTORS AND VOLUNTEERS

THE ARRESTED PROTECTORS:

Irene Abbey
Norman Abbey
Alison Acker
Tono Adams
Ralph Albert
Gabrielle Elizabeth Alden
Julia Allen
Barbara Celeu Amberton
Jesse Amberton
Hannah Amrhein
Craig V. Anderl
Cyndi Anderson
Gemma Lewis Anderson
Steven Anderson
Corrie Dee Archer
Brian Arnott
Erin Ashbee
Carl David Ashley
Ron Aspinall
Cam Atkinson
Oria Dawn Atkinson
Jacqui Lillian Aubuchon
Julie Austin
Corey Ross Avery
Jodi Ayers
Taylor Bea Bachrach
Kim Backs
Philip Bailey
James Gordon Baillie
Alex Baine
Riversong (Marion) Blaine
Christopher Adam Baker
Lorelei S. Baker
Zenia Faith Barbelas
John Barlow
Tamara April Barlow
Phineas Barnes
Craig Marshall Bartlett
Samadhi Bartock
Ruth Linda Bassingthwaighte
Kerrie-Lynn Bates
Kathleen Mary Battle
Emily Bauslough
Jennifer Yvonne Beaudry
Emma Tulameen Beaupre
Dean Begley
Jason Michael Bell
John Kevin Bell
Louise Ellen Bell
Warren Bell
Lloyd Thomas Bellaire
Eileen Bennett
Barbara Benoit
Len Bentson
Sara Louise Berger
Lorne David Berman
Tzeporah Berman
Wanda Best
Elaine Bilodeau
Karelyn Joy Bird
Lisa Blackwell
Laurie Allette Bloom
Lori Danielle Blumden
Rita Blunck
Davin Bobrow
Thilo Bode
Marcella Bodman
Monika Bogovic
Selina Jeannine Boily
Shirley Sarah Bonner
Rory James Booth
Shannon Michelle Booth
Faith Bordelas
Ria Boss
Jeremey Botly
Tim Boultbee
Geoffrey Bowers
John Robert Bowers
Jordan Bowers
Robin Jonathan Boxwell
Caroline Bradfield
Susanne Bradfield
Lyle Bradley
Stuart Bradley
Katy Brammer
Katie Brand
Laurel Brant
Phoebe Brant
Lutz Bräuer
Jessica Bristowe
Lauren Bristowe
Katie Brown
Terry Lee Brown
Graeme Brownlee
Catherine Bruhwiler
Mitzka Bryans
Ariel Bryers
Donald Buchanan
Marjorie Buchanan
Jamie-Lynne Burgess
Venessa Ruth Burgoon
Asha Burke
Mira Burke
Stanley Brian Burke
Neil Bradley Burnett
Heather Leanne Burney
Alexis Burton
Andrea Caddy
Sarah Calisto
Mira Burke
Stanley Brian Burke
Neil Bradley Burnett

Heather Leanne Burney
Alexis Burton
Andrea Caddy
Sarah Calisto
Richard Campbell
Maryanne Terese Campeau
Shawn Douglas Cantelon
Sean Gaelin Carriere
Sean Michael Carter
Christopher Eric Carteron-Ewington
Joan Cartwright
Persilia Caton
Frederick Ozmer Catt
Lindsay P. Caywood
Carole Chambers
Lyle Chambers
Nathan Michael Chambers
Petra Chambers
Brother Change
Ericka Jane Chemko
Vivian Chenard
Norman Damien Cholette
Niels Christensen
Linda Cirella
Kim Citton
Donald Clancy
Josh Clark
Amelia Clarke
Mike Clarkson
Samuel D. Clemens
Richard William Clements
Beverly Joy Clifford
Susan Devor Cogan
Adam Cohn
Diane Colins
Eroca Allison Colins
Wendy Collins
Suzanne Elizabeth Connell
Laura Connery
Paul Conningham

Vrinda Conroy
Michael Neil Conway
Parker Cook
Francois Corbeil
Rita Corcoran
Kevin Corrigan
Emile Coulter
Mikle Cowan
Dave Cowel
Brian Craigie
April Kathleen Cramer
Suzanne Crawford
Lendra Cristiano
Felix Cruz
Sandor Csepregi
Susan Culver
Paul John Cunningham
Jane H. Currelly
Judith Mary Currelly
Joseph Jean-Guy Custeau
Dave Cutts
Larissa Anne Czuchnowsky
Francis Daigle
Ryan Mead Dalrymple
Sharon Leslie Danroth
Karen Dauvries
Erin Catherine Davies
Mark Davies
Lance Davis
Alison E. Dawson
Karen Jane Dawson
Martine De Grandpre
Annette De Villiers
Arnoud DeBoer
Amy Alexandra deCarteret Feit
Kelli Lynne Deering
Martine Degranopre
Austin Delany
Bob Delgatty
Tom Joseph DeMarco
Jesse Demb
Jessica Naome Demers

Valeria DeRege-Thesaurd
James Patrick DesJardins
Gareth Patrick Devenish
Karen Devries
Mike Dewilde
Holly Virginia deWolfe
Norma J. Dewolfe
Nicholas Jay Diederichs
Melinda Diver
Kamala Ann Divers
Ron Dobie
Sandra Lee Dolph
Robert James Dolphin
Angela Donelly
Eric Donnelly
Casey Doss
Geert Drieman
Galena Dubeau
Elaine Marie Dubois
Michel Duhaime
Andrea Dumont
Rob Duncan
Luisa Durante
Cornelia Durrante
Laureen Jane Dutly
Joel Dyck
John Christian Egli
Hank Einarson
Jamie William Elder
Kristen Elder
Adriann Elkelenboom
Josine Elkelenboom
Elesa Elliot
Terra Nova Elson
Sandy Emerson
Jennifer Emile
Allen Leonard Engler
John David Enright
Jean Ensminger
Anja Louise Erichsen
Darren Erickson
Arianna Etheridge
Sarah Etheridge
Sasha Everett

Phyllis Fabbi
Roddy Raymond Fagan
Betty Fairbank
Anne Fairbanks
Maria Falerino
Aimee Falk
Daniel Falk
Stuart Fall
Claudia Fancello
Heather Edith Faris
Janes Fawkes
Patrick Fawkes
Gabriel Fernandez
Joanna Finch
Natalie Fisher
Marilyn Muriel Fitzmaurice
Danny Flanagan
Kellsie Forbes
Lisa Forbes
Mary Sand Ford
Cynthia Okie Foreman
Ray Forsberg
Marie Annie Fortin
Shane Fortune
Gwydndlyne Foster
Erica Foulkes
Patricia Francis
Patricia Fraser
Peggy Fraser
Susana Frazer
Liora Toba Freedman
Robin Freels
Jack Barry Friedman
Todd Allen Friedman
Andrew Brian Gage
Gavin Gagne
Gwen Gagne
Mary-Ann Gardias
Rayana K. Garen
Jean Pierre Gaudin
Marguerite Francine Gauthier
Christopher Gauvreau
Gwethalyn Gauvreau
Charlotte Geddes
Allison Gell
Sally Gellard
Ian Gemmell
Tania Christine Genoway
Philip Ralph George, Jr.
Li Lien Gibbons
Marguerite Yvonne Gibbons
Maurice Gibbons
Ben Gilbert
Lesley Jean Gilbert
Scotia Gilroy
Joanna Gislason
Patrick Gleahy
April Goebl
Mandu Goebl
Mike Goodliffe
Chris Randall Gorin
Mark Graham
Laurel Grant
Jennifer Green
Shane Leon Greene
Susan Figler Greenwood
Deborah Elizabeth Gregg
Elanne Grose
Heather Kristina Grove
Noah Grove
Elisa Charleen Guilbeault
Roger Guimond
Stefan Haerie
Brenda Mary Haggard
Jay Hamburger
Jim Hamilton
Patrick Hamilton
James Donald Hamly
Arne Boye Hansen
Linda Joann Hanson
Sue Hara
Steven McRae Harmer
Adam Harris
George Harris
John Scott Harris
Tyson Harris
Veronica Hartman
Gregory Paul Hartnell
Lisa Dawn Hartnup
Claire Marie Harvey
Darlene Mace Harvey
Hayden Harvey
Chris Hatch
Ronald Barrie Hatch
Robbyn Hatton
Klaus Hindrich
Hauschild
Annelies Haussler
Daniel Hawkes
Eric Hawthorne
Mary Hay
Angela Lynn Hayward
Bob Hayward
Keyo Lynn Hayward
Josh B. Heath
Ernest Hekkanen
Kirsten Helbig
Amy Janna Henderson
Drew Henderson
Paul Henderson
Nicolas Hervieux
Leesa Heyward
Shelley Hilbert
Allison Jean Hill
Amanda Hill
Victoria Margaret Hill
Susan Hjerpe
John Hodgson
Otto John Hoekstra
Carl Lawrence Hofbauer
Jackie Jane Holmes
Robert Holmes
Theresa Hood
Michael George Horn
Inaki Ardoni Houlihan
Stive Howard
Kenneth Howes
John Hubard
Todd Huffman

Ann Hughes
Ching France Hurst
Christopher Noa Hutchinson
Keither Wayne Jacklyn
Gail Jackson
Stuart Jamieson
James Jamly
Milo Jansus
Victor Vidutis Januska
Crystal Jennings
Leslie Arden Jennings
Bridget Jensen
Sara Jensen
Serena Faith Jewell
David Johanson
Valeria Patricia John
Carol Virginia Johnson
Geoff Johnson
Sidney Lorre Johnson
Charlie Johnston
Peter Andrew Johnston
Phillipa Joly
Redner Jones
Joan Kathleen Jubb
William Juby
Conrad Steward Juraschka
Shauna Kaendo
Dana Kagis
Mathew David Kagis
Susan Elizabeth Kammerzell
Evaline Karcher
Yvonne Marie Kato
Moira Veronica Keigher
Ralph Keller
Andrea Zoe Kellis
Malcolm M. Kemp
Erin Alida Kendall
Lauray Kendall
Pauline Kendall
Des Kennedy
Sandra Kennedy
Mike Kennett
Justin Kenrick
Victoria Kent
Benjamin Kersen
A. Khan
Deborah Michelle Kind
Patricia Ann King
Theron Paul Kingsley
Michelle Kinney
Steve George Kinnis
Michelle Kirshner
John Michael Kirkpatrick
Tom Kirwin
Dianna Lyn Klapkiw
Betty Chrystal Kleiman
Patricia Klein
Aragorn Klockars
David Knight
Ayala Monique Knott
Jennifer McIntyre Knowlan
Jennifer Koenen
Robert Larry Konkle
John Kortuem
Anita Josie Krajnc
Susan M. Krajnc
Betty Krawczyk
Moira Kreiger
Kevin Krogan
Quentin James Krogstad
Inger Kronseth
Ashley Kruse
Jan Kruse
Barbara Beata Kubicka
Grzegorz Andrzej Kubicka
Karlie Laboucan
Jannette Lade
Trevor David Lafond
Kim Elaine Lamont
Melissa Lampman
Heath Landsdowne
Karen Janet Lang
John Richard Laning
Jason Patrick Lanthier
Andrew Claudleigh Larcombe
Brandon Lau
Linda Margaret Laushway
Tony Law
Patrick Leahy
Tracy Leahy
Robert Lebaron
Edward Topp Leckie
Melanie Leduc
Amanda Lorien Lee
Harris Lee
Serena Lee
Vanessa Lee
Miriam Leigh
Nick Lenoire
Darlene Lessard
Gregory Letain
Vicki Levine
John Lewin
Chantelle Liboiron
Randy Liboiron
Peter James Light
Robert Light
Shad Light
Owen Lightly
Sheldon James Lipsey
John Victor Lironi
Jennifer Hallock Lodge
Anthony Ross Loring
Matthew William Lowe
Rob Lucas
Christine Lummis
Andrew MacDonald
Stuart Charles MacDonald
Zoe MacDonald
Keith Derek MacHale
Carol MacIsaac
Elizabeth MacKay
Naomi MacKay
Rebekah MacKay

Aaron MacKee
Hilary Mackey
Tara MacLean
Emily Anne MacNair
Katherine Alexandra Macy
Mary Madien
Willow Madill
Mary Madsen
Kim Magee
Robert Joseph Maher
Dr. Louis Patrick Maingon
Lea Ann Mallet
Mark Roger Mallet
Patrick Jeremy Mallet
Catherine Maneker
Joanne Charmian Manley
Keith Thomas Mansfield
Marc Justin Ryan Manson
Joseph Steven Marcotte
Ian Marcuse
Christian David Marr
Danae Marshalsay
Bernard Anthony Martin
Richard Martin
Robert William Martin
Samantha Martin-Evans
Shaye Martirano
Taylor Aaron William Martyn
Yone Massie
Herbert Damon Mathews
Ali Cynthia Mathias
David Bruce Matson
John Christopher Matson
Jennifer Maxwell
Melanie Maxwell
Bryan Maxwell
Justin James May
Ed May
Shannon Mayo
Leaf McClary
Linda McDaniels
Sean Rohan McDonald
William Robert McDougall
Gregg Vincent McElroy
Kim Marlane McElroy
Scot Buchanan McFadyen
Scott William McGee
Edward McGovern
Douglas Kenneth McGregor
Janet Ellen McIntyre
Brian McKean
Dariuys Alexander McKenzie
Jean McLaren
Tia McLennan
Joseph McNamee
Lee McNamee
Simon McNeill
Sharon McNulty
Dougal Murray Meekison
Luz Meyer
Jessica Jane Michalofsky
Amanda Miller
David Miller
Doug Miller
Heather Miller
James Edward Miller
Juanita Miller
Lorna Miller
Marilin Jo Miller
Eric Charles Milner
Constantin Mi-rochnitchenko
Jeremy Mitchell
Kenneth Metro Mitchell
Ken Molloy
Galen Mongeau
Heather Alix Montgomery
Niamh Moore
Fiona Moorhead
Faith Moosang
Donna Morgan
Philip Morgan
Chris Morrison
Sean Thomas Patrick Morrison
Courtland James Morrow
Darrin Mortson
Julie Mottram
Cora Mueller
Michelle Mueller
Esther Muirhead
Viola Müller
Jennifer Leigh Mundell
Gordon Michael Murphy
John Murphy
Kevin Murphy
Lisa Murphy
Tracey Myers
Gordon Myskow
Deepa Vasudevan Nair
Lesley Maryn Neilson
Duane Christopher Nelles
Martin Aaron Nelson
Jason Sebastian Nelsons
Mischa Neuman
Ervin Leonard Newcombe
Pearl Newfield
Charles David Nickarz
David Andrew Nicoll
Ian Taylor Nikolaus
Lisa Elaine Noel
Judy Norbury
Ana-Maria Noronha
Emily Nyberg
Myranda O'Byrne
Morgan O'Donnell
Patrick O'Rourke
Richard William Ogilvis
Darlene Kay Olesko
Jason Paul Oliver
Carl Benedict Olsen
Grant Mathew Olsen
Stephanie Olsen
Margaret Inessa Ormond

Bob Osleeb
John Val Osler
Carolyn Oyarzbel
Erica Ozemko
Andrew Carson Palmer
Georgia Paquette
Kim Paquette
Talia Paramio
Edward Robert Parker
Stuart Parker
Shawn Parkinson
Michelle Parks
Parmita Cara Parks
Anne Pask-Wilkinson
Donna Passmore
Michael Alan Passoff
Ralph Tom Paul
Kirsten Paxton-Judge
William Payme
Joseph Pearce
Marion Pearce
Melissa Pearce
William Pegg
Wolfgang Pekny
Summer Joan Pemberton
Kenneth Myron Penner
Robert Penny
Chris Peterson
Craig Peterson
Shawn Charles Peterson
Jesse Philips
Chris Phillips
Jane Savile Phillips
Wanda Phillips
Dilan Pickersgill
Jordana Pine
Jessica Plachta
Avan Plank
Wendy Pope
Barbara Porter
George Potvin
J'Aileen Pounder
Nancy Powell
Jeremy Preece
Jason Prefontaine
Elizabeth Susan Prentice
Daniel Prest
Susan Price
Eric Jonathan Priest
Christopher Thorton Pulford
Jonathan Pulker
Anne Francis Pyke
Dagen Quinn
Mara Quinn
Michelle Quinn
Rin Raap
Kalus Rabel
Sofia Raginsky
Paul Ralph
Susan Ramsay
Jean Breody Randall
Dean Rands
Mary Jean Rands
Yvon Alain Raoul
Sandy Ratcliffe
Mary Ann Rattray
Sheila Margaret Ray
Joe Rea
Lorraine Gail Reed
Tosh Reed
Aaron Rees
David Hugh Rees
Allan Reford
Mikhail Reid
Kelly Reinhardt
Nick Reynolds
Shawn Eric Rezansoff
Louis Joseph Rheame
Todd Andrew Richer
Tracy Rideout
Sara Jane Rider
Bernadette Ridley-Foster
Kathy Rieder
Colleen Lisa Riley
Mark Robert
Sam Roberts
Brian Douglas Robertson
Judith Aren Robinson
Rachel Nadine Rocco
Colin Donald Rolls
Peter Jackson Rowan
Nils Bernard Roy
Terrance Russel
John David Rutledge
Alison Samis
Donald Samis
Neil Saunders
Christopher John Savery
Jane Savile
Chelsea Saxe
Markus Saxienger
Gail Schacter
Christopher Cory Schile
Christine Schlattner
Joann Schmidt
Margret Schmidt
Mark Schroeder
Rye Schroeder
Heffa Schücking
Manuel D. Schulte
Oliver Schulte
Rayk Schuster
Christopher Thomas Scott
Ian Scott
Peter Gordon Scott
Davida Sefireth
Jennifer Leanne Serbu
Dan Sereda
Rya Naiome Shankman
Teresa Shanks
Stephanie Sharkey
Heather Fawn Shaw
Cory Sheedy
Tara Shelley
Seth Shugar
Muriel Sibley
Joseph Siddiqi
Miriam (Starhawk) Simos
Sile Simpson
Raelene Siobhain-Young
Aurum Rudy Skiare

Derek Smith
Dylan Robert Smith
Judy Smith
Kevin Douglas Smith
Steve Smith
Robert Smith-Jones
Layla Sneade
Florian Sommer
Daniella Sorrentino
Daniel Soundheim
Eileen Margaret Sowerby
Willy Sport
Chris (Brother) Springwater
Marie Natalie St. Amant
Carrie Louise St. Pierre
Layla Stavroff
Douglas Steary
Jakob Steele
Andrew Stevenson
Christian Stevenson
Marie Allison Stevenson
Katy Jane Steward
Cynthia Joan Stillwell
Noah Stillwell
Karen Stones
Anna Strunecki
Bradley Lawrence Stuart
Jennifer Leanne Stuart
Tammy Lynn Sullivan
Jan Svoboda
Andrew Swain
Oliver Swain
Dale Robert Swann
David Swanson
Shaan Syed
Jane Talbot
Alan Tatro
Jennifer Tatro
Stephanie Tatro
Timothy Frank Taylor
William Cameron Taylor
Heidi Teel
Leo Teta
Marc Evan Tetenman
Jill Danielle Thomas
Judith Aileen Thompson
Kevin Thompson
Linda Thompson
Robert James Thompson
Brook Thorburn
Karen Tinker
Janine Tomaney
Steven Trafideo
Charles Truscott
Peter Gwyn Tucker
Gerrard Tyler
Dale Ulrich
Frank Ute
Chandra Van Esterick
Rients Johannes Van Goudoever
Jordan Vander Goes
Berna Vanderhaan
Marcus Chrisopher Vanderzon
Erin James VanFleet
John Andrew VanHal
Stephan Vanicek
John Vedova
Erik Volet
Zoe Rosemary Wade
Kellie Ann Walker
Marion Walker
Melissa Walker
Nicole Walker
Robert David Sean Walker
Ruth Walmsley
Emily Wareing
Robbie Warren
Gregory Warry
Kim Washburn
David Watson
Doone Watson
Kathryn Watson
Rick Watson
Susan Elizabeth Jo Watson
Tim Watson
Paul Watt
Michael David Webb
Terry Gordon Weber
Tashia Weeks
Sara Wees
Timothy John Wees
Agnes Krystyna Wejman
Hugh Wellman
Deryk Wenaus
Guy Wera
Kristen Werner
John Allen West
Andrew Wheatley
Sue Wheeler
Andrew Cameron White
John White
Lalania Enid White
Tanis Whittaker
Anne Wilkinson
Loren Wilkinson
Mary Ruth Wilkinson
Merv Wilkinson
James Sidney Willer
Christopher David Williams
Cindy Williams
Cynthia Williams
Ole Williamson
Pete Willson
Daphne Wilson
Diana Wilson
Janice Wilton
Shelley Wine
Kelly Winkenhower
Robert Winkenhower
Gabriela Wolf
Ernestine Wood
Frances Louise Woodcock
Irene Leona Woodin
Donovan Woollard
Martyn Gregory Woolley
Richard Michael Woolsey

John Wright
Johnna Wright
Chris Wyman
Andy Yakimishyn
Carol Oyar Zabal
John Zelinski

SOME WHO HELPED AT THE FRIENDS OF CLAYOQUOT RESOURCE CENTER IN VICTORIA:

Bobby Arbess—employee
Jane Calder—employee
Claudia Cole—employee
Deirdre Gotto—employee
Edward May—employee
Scott McFadden—employee
Gill Thomas—employee
Gen Saini—employee
Mac Scott—employee
Andy Wheatley—employee

Alison Acker
Susan Arsaoon
Ria Booo
Janine Bandcroft
Jan Bate
Janie Benisher
John Bowers
Shannon Buchan
Vanessa Burgoon
Billy Burnstick
Eva Caputa
Jane Carrico
Amy Cassidy
Conky Conconan
Suzanne Connell
Sandor Csepregi
Jane Curelly
Mike Darney
Cameron Delacroix
Joan Deturbeville
Gareth Devenish
Eric Doherty
Luisa Durnate
Fred Easton—lawyer
Tom Feakens
Tom Forbes
Chris Forrest
Peggy Fraser
Mary Graham
Heather Grove
Noah Grove
Joan Gubb
Sarah Harris
Derek Hawksley
Ken Hough
Marty Hykin
Leslie Jennings
Dana Kagis
Paula Kahn
Jody Keeping
Victoria Kent
Ayala Knott
Inger Kronseth
Heath Lansdowne
Barbara Lee
Dan Osleeb
Tara Maclean
William Madill
Gordon McAllister
Keith McHale
Glen Milbury
Francis Moore
Terry Moore
Carol Mowat
Doreen Mueller
Alex Mungall
Jenny Murray
Michael Neate
Dave Nickarz
Colleen O'Donnell
Saki Ola
Lynn Pattison
Amanda Pedrick
Gene Phillips
Wanda Phillips
Steve Porter
Al Refond
Louis Rhaeume
Gail Schacter
Elizabeth Schmidt
Olga Schwantzcopp
Genni Sharman
Barbara Scott
Noel Scott
Eric Smith
Kriotal Stetsko
Carrie St. Pierre
Ingrid Strunzel
Sybil
Daphne and Hugh Taylor
Fran Thoburn
Lisa Tyler
Chip Vinnai
Jo Vipond
Deryk Wenous
John White
Fran Woodcock
Zebo

LAW STUDENTS WHO HELPED WITH APPEAL AND BAIL DOCUMENTATION:

Ms. Dawn Goodwin
Ms. Tamara Golinsky
Ms. Melaine Murray
Mr. Gerry Walsh
Ms. Shannon Williams

RECOMMENDED READING

Boyle, T.J.B. Biodiversity of Canadian forests: current status and future challenges. *Forestry Chronicle* 68: 444-453.

Ledig, F.T., 1993. Secret extinctions: the loss of genetic diversity in forest ecosystems, pp. 127-140 in M.A. Fenger et al. (eds.), *Our Living Legacy.* Proceedings of a Symposium on Biological Diversity, Victoria, B.C.

Pojar, J., 1993. Terrestrial diversity of British Columbia, pp 177-190 in M.A. Fenger et al. (eds.), *Our Living Legacy.* Proceedings of a Symposium on Biological Diversity, Victoria, B.C.

Wilson, E. O., 1988. The current state of biological diversity, pp 3-18 in E.O. Wilson (ed.), *Biodiversity*. Washington, D.C.: National Academy Press.

Alpert, B., 1993. "Coalition may be thorn in the side of environmentalists." San Diego Bail Transcript, Vol. 108, No. 129, p. 18.

Canada Library of Parliament, 1992. *Share Groups in British Columbia.* Ottawa: Canadian Government.

Capra, F., 1968. *The Turning Point*. New York: Bantam Books.

The Cato Institute, 1993. *Advancing Civil Society*. Washington D.C.

Deal, C., 1993. *The Greenpeace Guide to Anti-Environmental Organizations*. Berkeley: Odonian Press.

Devall, W., ed., 1993. *Clearcut: The Tragedy of Industrial Forestry*. San Francisco: Sierra Club Books & Earth Island Press.

Drushka, K., B. Nixon and R. Travers, eds., 1993. *Touch Wood: BC Forests at the Crossroads*. Madeira Park: Harbour.

Ehrenfeld, D., 1981. *The Arrogance of Humanism*. Oxford University Press.

Goldberg, K., 1991. Clearcuts, Moonies and Right-wing Death Squads. *Canadian Dimension*, July-August, p. 20.

Gottlieb, A.M., 1989. *The Wise Use Agenda: The citizens policy guide to environmental resource issues*. Bellevue: The Free Enterprise Press.

Hager, M., 1993. Enter the Contrarians. *Tomorrow.* Vol. 3, No. 4, pp. 10-19.

Hammond, H., 1991. *Seeing the Forest among the Trees: The Case for Wholistic Forest Use*. Winlaw: Polestar.

Institute of Contemporary Studies, 1993. International Center for Economic Growth Prospectus. San Francisco: ICS.

Jefferson, R., 1991. Touchy-feely environmentalists ignore science, costs and freedom in their crusade to run the world. *Trilogy*, May-June, pp. 48-51.

Loomis, R. with M. Wilkinson, 1990. *Wildwood: A Forest for the Future*. Gabriola Island: Reflections.

Miller, T., 1990. *Living in the Environment*. Belmont: Wadsworth, Inc.

National Legal Center for the Public Interest, 1993. *Annual Report*. Washington: NLCPI.

O'Keefe, M. and K. Daley, 1993. Ecofeature: Checking the right buzzword. *The Environmental Journal*, Vol. 5, No. 3, pp. 38-44.

Pacific Legal Foundation, 1993. *1992-1993 Annual Report*. Sacramento: PLF.

Postrel, V.I., 1990. The Green Road to Serfdom. *The Reason Environmental Reader*, April, pp. 2-8.

Satchell, M., 1991. Any color but green: A new political alliance is battling the environmental movement. *U.S. News and World Report*, October, pp. 74-76.

GLOSSARY

backspar trail: path for backspar equipment—the rigging necessary to hold the furthest spar in place

biodiversity: diversity of plants, animals and other organisms in all their forms and levels of organization, including genes, species, ecosystems, and the evolutionary and functional processes that connect them

bucking: cutting a felled tree into specified lengths

cants: large-dimension, rough-cut lengths of wood, often for export

clearcutting: felling and removal of all trees in an area

CORE: Commission on Resources and Environment, an attempt at land use planning in B.C., 1993-94

dimensional lumber: lumber cut into pieces ready for the construction trade

faller: the timber worker who falls trees

feller-buncher: machine used to fell trees by slicing them off near their base; some also strip limbs and gather logs in bunches (this most gruesome of machines has replaced many on-the-ground forest workers)

Forest Management Licence: form of forest tenure prior to the current Timber Supply Area system

Forest Practices Code: legislation, standards and regulations governing forest management in B.C.

grapple-yarder: a machine for hauling cut trees out of the bush

high-grading: removal of the best trees from a stand, often leading to a stand of poor quality trees

leave strips: strips of forest next to waterways that are required to be left uncut

log scaler: forest worker who estimates the size and volume of standing timber prior to its being cut

long line logging rigs: equipment used for hauling cut logs out of the forest involving one or two spar-trees with a long line in between, going a considerable depth into the woods

mens rea: state of mind as to guilt or innocence

monoculture: growth of a single crop, usually of the same age and species

NDP: New Democratic Party (the party in power in B.C. during the Clayoquot crisis)

NSR: Not Satisfactorily Restocked: productive forest that has been stripped of trees and has failed, partly or totally, to regenerate, either through natural means or by replanting

old-growth forests: a mature forest ecosystem containing a broad diversity of plant and animal species, and relatively uninfluenced by people

PAS: Protected Area Strategy—a government program which is supposed to identify areas to be saved from logging

PHSP: Pre-harvest Silvicultural Prescription, a process for gathering field information, setting management goals and standards for silviculture, and prescribing ways to meet these goals and standards

Raging Grannies: groups of grandmothers prone to wearing hats festooned with flowers, inventing and performing publicly outrageous songs set to the tunes of old standards in protest for environmental protection and peace

regeneration: growth of new forest after cutting, either through natural reseeding or artificially by replanting

riparian zones: areas next to waterways

RPF: Registered Professional Forester

second-growth: forest that grows after logging of old-growth

selective logging: cutting of only the most valuable trees, value being measured by species, size or quality

Share groups: timber industry-oriented and -sponsored groups in Canada, modelled after the same groups in the U.S.—called Wise Use groups there

silviculture: theory and practice of producing and tending a forest; controlling the establishment, composition, growth and quality of forest stands to achieve management objectives

skid trail: trail used by skidding machinery to move trees to a landing, ready for trucking away

skidding: sliding and dragging logs from where they are felled to a landing

skyline logging: a method involving the use of a spar (a natural tree, or a portable steel spar) to bring timber off a mountain instead of building roads

SLAPPs: Strategic Lawsuits Against Public Participation—used by industry and government to attempt to intimidate citizens with legitimate grievances into silence

snags: standing dead tree from which leaves and most branches have fallen

Socred: short form for the Social Credit Party of B.C., which held power for many years prior to the election of the NDP in 1991

spars: long poles rigged for dragging logs out of the woods in high-rigging or long-line logging, or for unloading logs from a truck or train at a log dump

stumpage: price paid to provincial government for timber cut on Crown land

temperate forest: one of three main forest types in the world—the other two types being tropical evergreen and northern coniferous forest—consisting mostly of deciduous trees (whose leaves drop in winter)

TFL: Tree Farm Licence—tenure agreement giving exclusive timber harvesting rights and management responsibilities in an area, on Crown and private land

timber licence: licence to cut and remove Crown timber

towers: a frame for a large grapple-yarder, or a tower on which is mounted a fire look-out

value-added production: manufacturing that adds value to raw wood as it passes through processing; measures profits against wages

watershed: area of land that drains into a waterway

(Sources: B.C. government, *British Columbia Forest Practices Code*; Forestry Canada, *The State of Canada's Forests 1992*; Forestry Canada, *The State of Canada's Forests 1993*; Merv Wilkinson....)

PHOTOGRAPH DESCRIPTIONS

Page ii: Demonstration at the opening of the British Columbia Legislature, March 3, 1993 — Lawrence McLagan.

Page xvi: Meares Island, Clayoquot Sound— Adrian Dorst.

Page 18: Clearcutting at Hesquiat, Clayoquot Sound — Adrian Dorst.

Page 48: Arrests at the Clayoquot blockade, August 1993 — Lawrence McLagan.

Page 64: The Peace Camp at the infamous Black Hole, August 1993 — Lawrence McLagan.

Page 120: Peter Garrett of the rock band Midnight Oil addresses protectors at the blockade, July 1993 — Lawrence McLagan.

Page 160: Vigil outside the Law Courts in Victoria, B.C. — Chip Vinai.

Page 178: The daily blockade at Kennedy Lake bridge, Clayoquot Sound, July 1993 — Lawrence McLagan.

Page 188: Dawn at the blockade at Kennedy Lake bridge, August 1993 — Lawrence McLagan.